© 2011 Lulu Author. Dorota M. Goede B.Arch. All rights reserved.

ISBN 978-1-257-94657-0

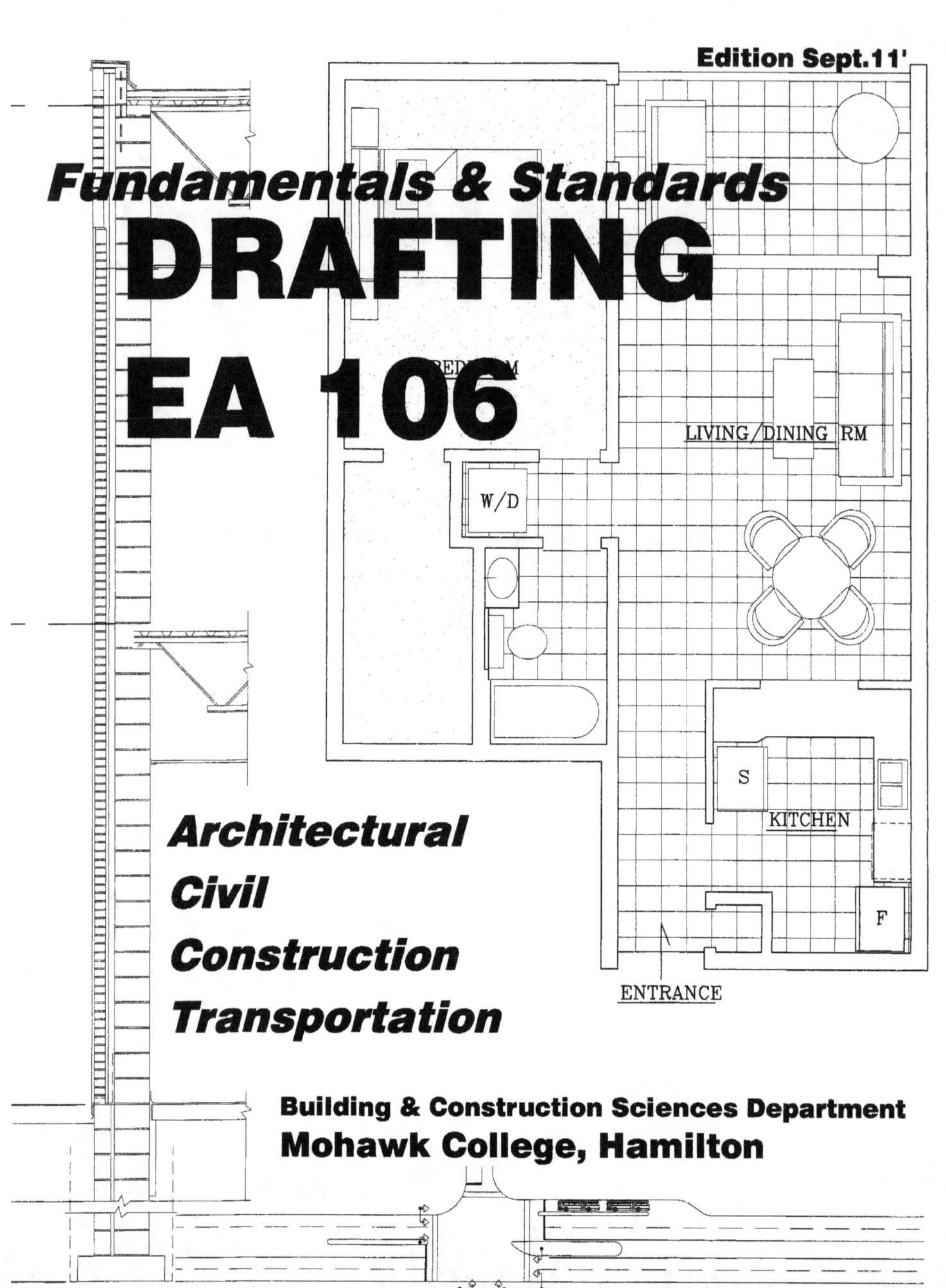

Edition Sept.11'

Fundamentals & Standards
DRAFTING
EA 106

Architectural
Civil
Construction
Transportation

Building & Construction Sciences Department
Mohawk College, Hamilton

EA106 REQUIRED EQUIPMENT LIST

ARCHITECTURAL – CIVIL – TRANSPORTATION CLUSTER

Students should arrive at their first drafting class in the first week of classes equipped with the following drafting equipment (minimum requirement):

Bookstore #:

1. Introduction to Drafting EA 106 Course Notes, latest revision
2. 36" T – Square and case (detachable head optional)
3. 12" 30/60/90 Set Square (drafting triangle) — 964 12-60
    18" Set Square is recommended for convenience with long lines!
4. 10" 45/45/90 Set Square (drafting triangle) — 964 10-45

Architectural Students

5. Imperial architectural scale rules (triangular)
    3/32", 1/8", 3/16", 1/4", 1/2", 3/4 ", 1", 1 1/2", 3" — 987 18-31
6. Metric architectural scale rule (triangular)
    1:20, 1:25, 1:50, 1:75, 1:100, 1:125 — 987 18-01

Civil – Construction – Transportation Students

5. Imperial architectural scale rune (triangular)
    3/32", 1/8", 3/16", 1/4", 1/2", 3/4 ", 1", 1 1/2", 3" — 987 18-31
6. Metric engineer's scale rule (triangular)
    1:100, 1:200, 1:250, 1:300, 1:400, 1:500 — 987 18-04

Important Note: Do not purchase scale with just numbers on sides: 16, 20 etc.

7. Mechanical pencil (thick-lead for drafting, thin-lead for notes only!)
    H and 2H pencil leads; do not use HB for drafting! — at the cash register.
8. Drafting Pencil Lead Sharpener; Mars Lead Pointer & Ziplock Bag! — Keep Clean !!!
    031901905699
9. White plastic eraser and erasing shield
10. Lettering Guide
11. Masking Tape (the best is green painter's tape readily available at CTire or HDepot.
    Please note that the small drafting tape sold at the bookstore does not work well),
12. 4" dia., Drawing Tube
13. Drafting Board Cleaning Brush
14. Drawing cleaning pouch bag filled with eraser shavings
15. Drafting Paper: A1, A2 size **transparent vellum**;
    Do not use bond paper for drafting A4 smallest sheets will also be used but can be obtained by cutting A2 to 4 equal parts
16. 6" Compass set, Circle Template,
17. Calculator as required for other courses; simple calculations are required for drafting
18. Fine Line black Marker and HB pencil lead fro freehand drawings
19. French curve set, or flexible plastic curve (short)

NOTE: Drafting room availability is somewhat limited outside class time. Students may wish to consider the purchase of their own drafting table to use at home.
All above-mentioned equipment is available at Mohawk College bookstore.

IMPORTANT ADDITIONAL NOTES:

- Architectural students should purchase residential furniture and appliance template containing 1:100 and 1:200 scales. This template can be used for ¼" and 1/8" imperial scales.
- All students should have added by now a simple circle template; template containing other geometric shapes will be of a great value.
- If smudging was your problem last semester, purchase an eraser pouch used to clean drafting vellums of pencil dust.
- Drafting room availability is somewhat limited outside class time. Students may wish to consider the purchase of their own drafting table to use at home.
  All above-mentioned equipment is available at Mohawk College bookstore.

INTRODUCTION

Welcome to Mohawk College! The faculty of the Building and Construction Sciences Department wishes you success and fun while learning the exciting professions of ideas and planning, intelligent design, drafting, detailing and construction.

The primary aim of this particular manual is to provide you, and your instructor, with practical drafting exercises vital to the field of architectural, civil, transportation, and construction technology.
Exercises include preliminary lettering, drafting and geometry construction techniques, followed by more advanced field specific sections. You will be exposed to all disciplines taught in our department. Should you decide that your interests lie in field other than the one you are currently enrolled in, you will be able to apply for a transfer in programs at the end of the first semester.

The manual is a collaborative effort of our department faculty. As we keep revising and improving the content, your input will be appreciated. Discuss with your instructor any suggestions you may have.

Keep in mind that the design and drafting of any building or structure is not easy. Even experienced professionals must work hard and long and they will always have their moments of doubt. The rewards come with experience, patience and practice. Good luck in your program of studies and in all your future endeavors.

We are introducing a course website with this year's edition. Please make sure you take advantage of many valuable references and postings to supplement your learning in the classroom. Refer to the College's Mocomotion website and access "My Courses" AND THEN "The Consolidated Courses" page at the right top corner!. Select Introduction to Drafting page from presented options. This will be your generic information site. Your instructor may use the Introduction to Drafting webpage under your current semester "My Courses" page. The information presented here will differ. Get into habit of checking both sites regularly.

Dorota M. Goede B.Arch., LEED Green Assoc.
on behalf of **Faculty and Staff**
Building and Construction Sciences Department   September 2011

Mohawk College
HAMILTON   ONTARIO

INTRODUCTION TO DRAFTING
NAME: Dorota M. Goede B.Arch

CLASS: ALL
DATE: Sep.07'

DWG 00

MANUAL CONTENT and OBJECTIVES

MODULE 1 – Preliminary Drawing Techniques

OBJECTIVE
Upon successful completion of this module, the student will have demonstrated an ability to recognize and understand orthographic and isometric visualization techniques, and will have demonstrated an ability to comply with established standards for lettering, linework, and dimensioning for the architectural, civil/construction, and transportation sections of the course of study.

MODULE NUMBER	DETAILED OBJECTIVES	RESOURCES	TIME (HRS)
1.1	SCALES: Demonstrate correct and accurate use of Imperial and metric scale rules.	Instructor Manual	3
1.2	LINE DRAFTING EXERCISES: Demonstrate an ability to correctly and clearly produce linework assignments of an acceptable quality.	Instructor Manual	3
1.3	TECHNICAL HAND PRINTING: Demonstrate ability to correctly and clearly produce hand lettering of an acceptable quality.	Instructor Manual	3
1.4	BUILDING PERIMETERS: Demonstrate ability to draw and calculate building perimeters and areas.	Instructor Manual Estimating	3
1.5	GEOMETRIC CONSTRUCTIONS: Demonstrate ability to understand, and manipulate the basic principles and applications of geometry into hand-sketched and formal drawing styles.	Instructor Manual Reference	3
1.6	FREEHAND DRAWING PRACTICE: Demonstrate ability to draw geometrical shapes freehand for application in detailing and designing.	Instructor Manual Reference	3
1.7	ORTHOGRAPHIC PROJECTIONS: Demonstrate ability to understand, and manipulate the basic principles and applications of orthographic projections into hand-sketched and formal drawing styles.	Instructor Manual Reference	3

TOTAL: 21 hrs

MODULE 2 – Architectural Drawings

OBJECTIVE
Upon successful completion of this module the student will have demonstrated an ability to meet departmental drafting standards by successfully completing a selection of architectural, civil, and transportation drawing assignments.

MODULE NUMBER	DETAILED OBJECTIVES	RESOURCES	TIME (HRS)
2.1	ADVANCED FREEHAND DRAWING: Demonstrate an ability to prepare freehand perspective drawings of geometric compositions, construction details, and existing buildings.	Manual Instructor References	3
2.2	ADVANCED GEOMETRY: Demonstrate an ability to draw varied angle layouts and complex geometry using available drafting tools.	Ditto	3
2.3	LANDSCAPING PROJECTS: Demonstrate an ability to design and draw simple residential outdoor structure projects. Calculate estimates for materials and material lists.	Ditto	3
2.4	FURNITURE AND FIXTURE LAYOUTS: Demonstrate an ability to design and draw simple furniture and fixture layouts using typical professional office techniques.	Ditto	3
2.5	HOUSES: Demonstrate ability to draw residential Site and Floor Plans and Elevations. Prepare drawings of a residential project from materials as specified by the instructor. Demonstrate an ability to draw simple residential construction details.	Ditto	6
2.6	CONSTRUCTION DETAILS: Demonstrate an ability to prepare a civil/construction oriented drawing assignment. Demonstrate an ability to prepare transportation- oriented drawing assignment.	Ditto	6
		TOTAL:	21 hrs

MODULE 3 – Civil & Construction Drawings

OBJECTIVE
Upon successful completion of this module, the student will have demonstrated basic drafting techniques required for civil & construction-oriented drafting.

MODULE NUMBER	DETAILED OBJECTIVES	RESOURCES	TIME (HRS)
3.1	**ADVANCED GEOMETRY:** Demonstrate an ability to draw varied angle layouts and complex geometry using available drafting tools.	Instructor Manual References	3
3.2	**FOUNDATION DETAILS:** Demonstrate an ability to draw civil and construction drawings showing footings, columns and foundations.	Ditto	3
3.3	**FOOTING REINFORCEMENT:** Demonstrate an ability to draw civil and construction drawings showing foundation reinforcement details.	Ditto	3
3.4	**STRUCTURAL PLANS AND SECTIONS:** Demonstrate the ability to draw civil & construction-related plans and sections for industrial/commercial facilities as directed by the instructor. Draw construction details for brickwork and blockwork.	Ditto	3
3.5	**SURVEYING PROFILES & CALCULATIONS:** Prepare survey-type drawings showing plan, profile, original ground line, municipal services, subdivision layout, and roadways. Perform calculation pertaining to cut and fill, trenching, areas, volumes, co-ordinates, and costs.	Ditto	3
3.6	**CONSTRUCTION DETAILS:** Demonstrate an ability to prepare a transportation-oriented detail drawing assignment.	Ditto	3
3.7	**CONSTRUCTION DETAILS:** Demonstrate an ability to prepare an architecture-oriented drawing assignment.	Ditto	
		TOTAL:	21 hrs

MODULE 4 – Transportation Drawings

OBJECTIVE
Upon successful completion of this module, the student will have demonstrated basic drafting techniques required for transportation-oriented drafting.

MODULE NUMBER	DETAILED OBJECTIVES	RESOURCES	TIME (HRS)
4.1	ADVANCED GEOMETRY: Demonstrate ability to draw varied-angle site plan layouts and complex geometry using available drafting tools. Prepare survey-type drawings showing plan, profile, original ground line, municipal services, subdivision layout, and roadways	Instructor Textbook	3
4.2	ARC CONSTRUCTIONS: Demonstrate ability to draw	Ditto	3
4.3	INTERSECTIONS: Demonstrate ability to draw transportation drawings showing simple road intersections.	Ditto	3
4.4	PRELIMINARY TRANSPORTATION DESIGN: Demonstrate the ability to draw plans, profiles, and cross-sections of roadway and off-road facilities.	Ditto	6
4.6	CONSTRUCTION DETAILS: Demonstrate an ability to prepare a civil/construction-oriented drawing assignment.	Ditto	3
4.7	Demonstrate an ability to prepare an architecture-oriented drawing assignment.	Ditto	3
		TOTAL:	21 hrs

MODULE 5 – Construction Engineering Technician Drawings

OBJECTIVE

Upon successful completion of this module, the student will have demonstrated basic drafting techniques required for transportation-oriented drafting.

MODULE NUMBER	DETAILED OBJECTIVES	RESOURCES	TIME (HRS)
5.1	ADVANCED GEOMETRY: Demonstrate ability to draw varied-angle plan layouts and complex geometry using available drafting tools.	Instructor Textbook	6
5.2	ADVANCED BLUEPRINT READING: Demonstrate ability to recognize and read the standard drafting techniques as being currently used in residential and commercial construction.	Ditto	6
5.3	CONSTRUCTION DRAWINGS: Demonstrate ability to research and draw construction-oriented drafting assignments.	Ditto	9
		TOTAL:	21 hrs

SCALES

Professionals use triangular or beveled scales according to their preference. Your kit includes architectural (imperial), and engineering (metric) scales. Designers and drafters in Canada use metric and imperial measure systems as requested by their Clients. You must become proficient in using both scales.

Basic rules to follow:

- Engineering, metric scale:

    1. Large-scale site plan and transportation drawings are drafted using "m" as a measure unit.

    2. Floor plans, elevations, sections and details are drafted using "mm" as a measure unit. All dimensions shown on drawing are millimeters but you must use scale showing only meters. This can be frustrating for the beginner drafter, as you have to constantly switch between meters and millimeters in your mind.

    3. 1m = 1000 mm, 1 meter = 1000 millimeters
       1mm = 0.0001m, 1 millimeter = 1/1000 meter

       1 dm = 100 mm
       1 cm = 10 mm

    4. All metric dimensions on architectural, civil, construction and transportation drawings should be rounded off to the nearest 0 or 5mm

    5. Typical scales used for drawings are:

       1:1000, 1:500, 1:250, 1:200   Site Plans
       1:200   Floor Plans for large buildings, Site sections
       1:100   Typical for Floor Plans, Building Elevations and Sections
       1:50   Floor Plans, Elevations and sections for small buildings
       1:20, 1:25   Wall Sections
       1:10, 1:5   Construction Details and Connections

SCALES – cont'd

Basic rules to follow – cont'd:

- Architectural, imperial scale:

1. There are really two architectural scale systems. One is based on full scale and proceeds to smaller scales; the other is based on 1/16"=1'-0" scale and proceeds to larger scales. Combined these two systems offer a wide range of scale choice.

Table A	Table B
12"=1' (full scale) Mech Details	1" = 1'-0" Details
6"=1'	½" = 1'-0" Wall Sections
3"=1'	¼" = 1'-0" Residential Floor Plans
1½"=1'	Elevations, Sections
¾"=1'	1/8" = 1'-0" Commercial Floor Plans
3/8"=1'	1/16" = 1'-0" Residential Site Plans
3/16"=1'	
3/32"=1'	1"=10' etc. Large Site Plans

2. As 1' = 12" you must be very careful not to use metric, decimal, additions.

3. Convert all fractions of an inch into a common denominator prior to performing any mathematical additions.

4. Use scale as follows.

    Table A: Set the required number of inches and fraction of inches on the first point mark on your drawing. Now follow the required number of full feet, or inches, on a scale. Make a required measurement point on your drawing.

    Table B: Set the required number of inches on the first point mark on your drawing. Now follow the required number of full feet on a scale. Make a required measurement point on your drawing.

5. Please note that you will be working with your scale from left to right or from right to left depending on the scale. Each imperial scale bar has two scales marked on it.

SCALE CONVERSIONS

The ability to calculate fractions is a very important skill used in drafting as almost all architectural, civil, construction and transportation drawings are drawn to a scale. Sizes of construction components must be transferred to an appropriate measurement.

Example of a real length calculation using an architect's ¼" scale calculation:

Question: How long is the line measuring 5 5/8" using the ¼" scale?

Solution: $\dfrac{5\ 5/8"}{1/4"}$

$5 \times 8/8 + 5/8 = 40/8 + 5/8 = 45/8$

$\dfrac{45/8}{1/4}$

$45/8 \times 4/1 = 45/2 \times 1/1 = 45/2 = 22\ 1/2$ intervals = 22 ½ feet
= 22'- 6"

Answer: The line measures 22'- 6"

ASSIGNMENT: Complete the following table by calculating the real lengths of lines using the appropriate scale:

Length: \ Scale:	½" = 1'-0"	¼" = 1'-0"	1/8" = 1'-0"
5 ½"			
7 ¾"			
10 3/8"			
5 3/16"			
7 11/32"			
14 8/16"			

Mohawk College
HAMILTON   ONTARIO

SCALE CONVERSIONS
ASSIGNMENT

DWG

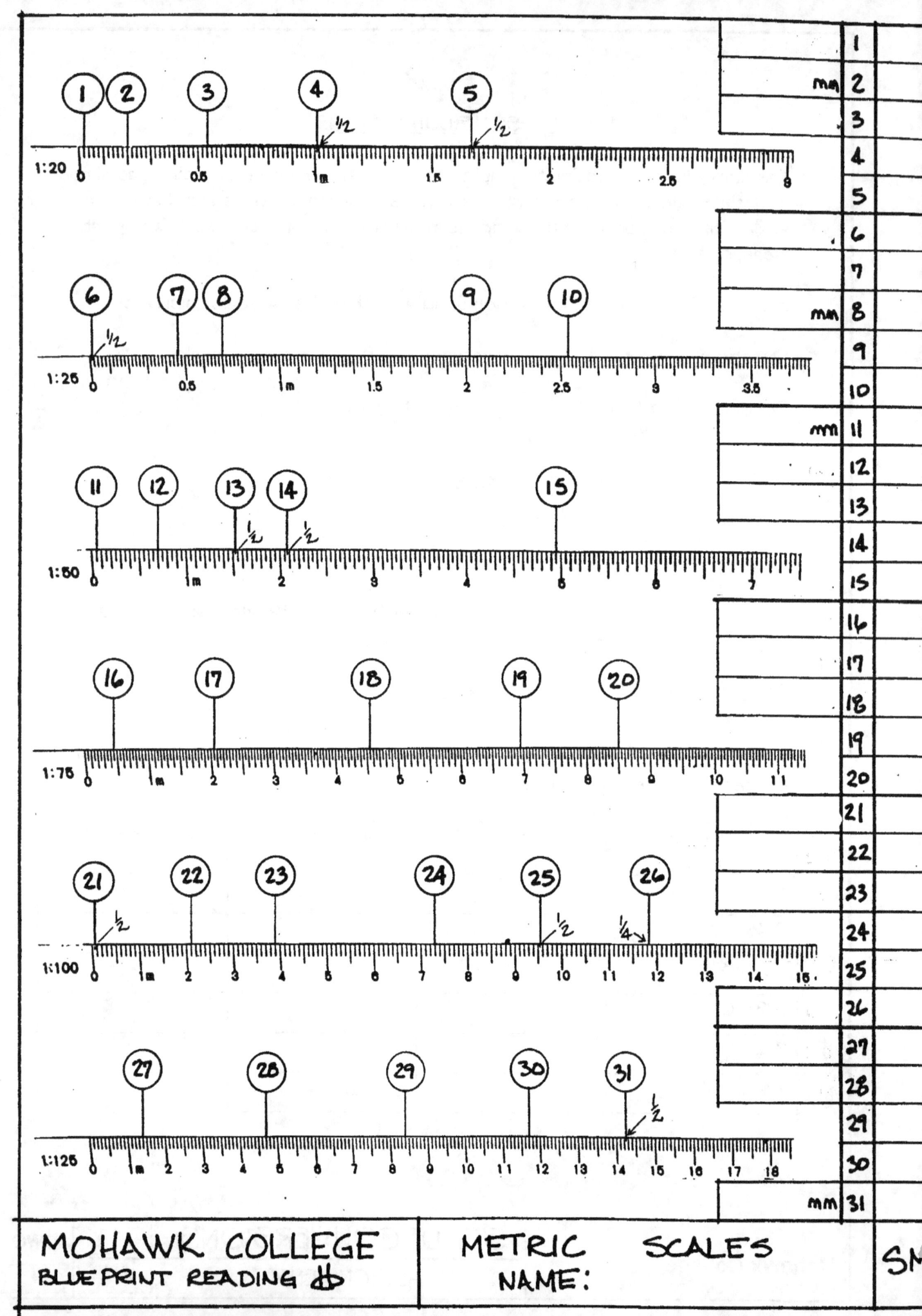

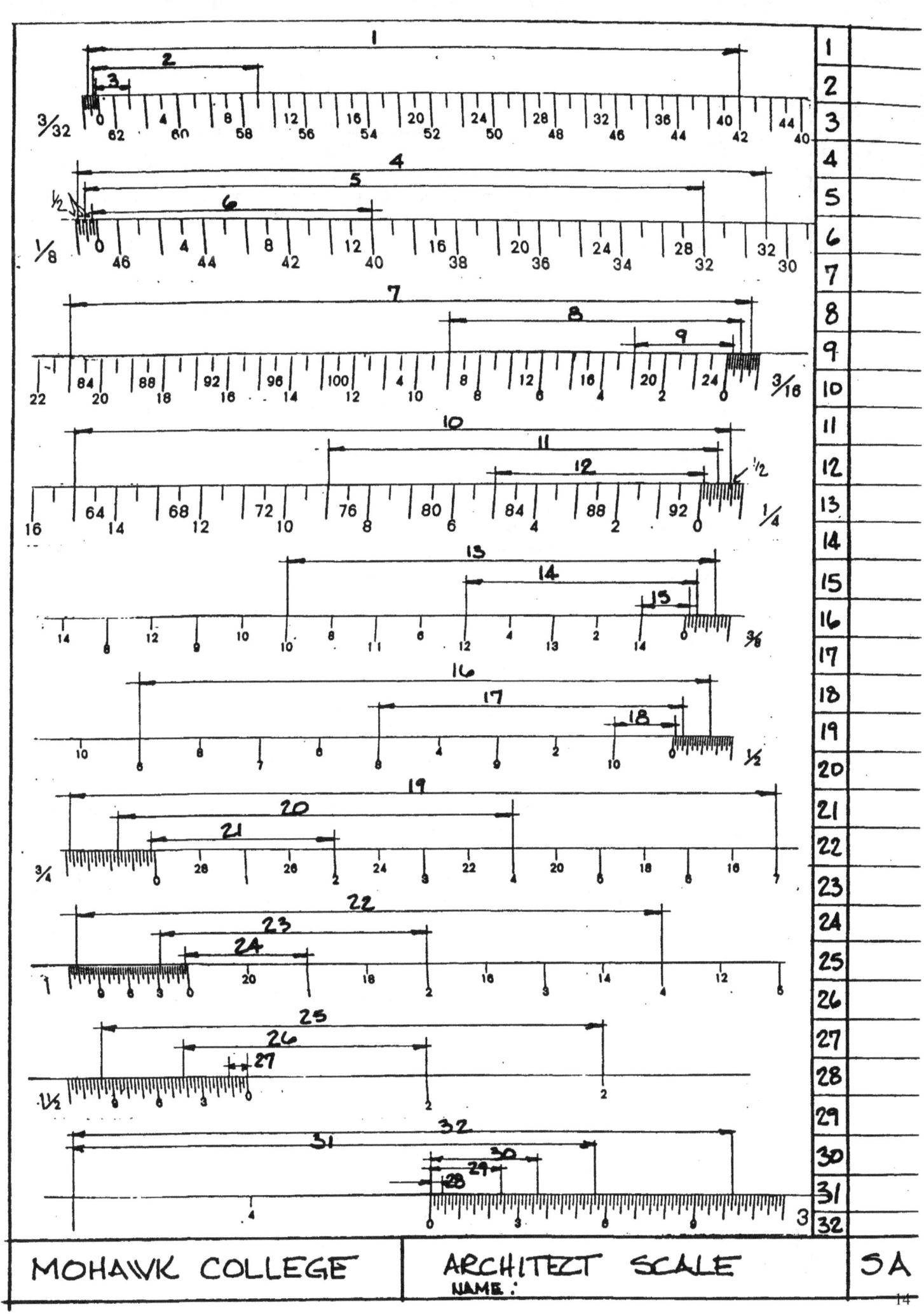

DRAFTING BASICS; LINE PRACTICE EXERCISE

Drafting is the term used to describe a task of producing drawings with the aid of special mechanical tools. The lines produced are straight and of constant weight. The list of basic drafting tools required for this course is attached at the very beginning of your manual.

Horizontal lines are drafted using the top edge of a T square, parallel rule, or a drafting machine. Vertical lines are drafted using the left edge of a triangle placed tightly against the T square. Most inclined lines are drawn from left to right using the top edge of a triangle after it has been set to the desired angle. A compass sharpened to a wedge point by sanding just one flat surface produces circles and arcs.
Your instructor will explain and show you how to use these basic drafting tools.

Sharpen your pencil. Draw all lines by constantly turning your pencil with your fingers. For drafting very long lines stop gradually by finishing with a lighter section. Continue the long line by starting over the just completed light section. Try to place the tip of your pencil exactly on the last line section. Gradually press harder and match the line weights.

LINE PRACTICE EXERCISE:

Use A2 size drafting vellum sheet. Refer to Appendix, at the end of this manual, for standard metric size sheets. Draw sheet margins, also per Appendix, and practice long line technique described above. Margins are always drafted very dark. The larger left hand side margin allows for future binding of blueprints into sets.
The objective of the line exercise on the next page is to practice drafting lines from no pressure on pencil whatsoever to the maximum possible weight. The light line is drafted using only the pencil weight; the darkest one is drafted by applying lots of pressure and by turning the pencil continuously. Start with offset lines from margins and establish sheet layout by placing light construction lines. The square can be started from a centrally placed point within the allowed drafting space.
The lightest lines shown are construction lines drafted first in order to compose the sheet layout. The square construction lines are also necessary for the accurate geometry.

Dorota Goede B.Arch

Mohawk College HAMILTON ONTARIO	LINE DRAFTING	CLASS	DWG
	LINE PRACTICE EXERCISE	DATE	

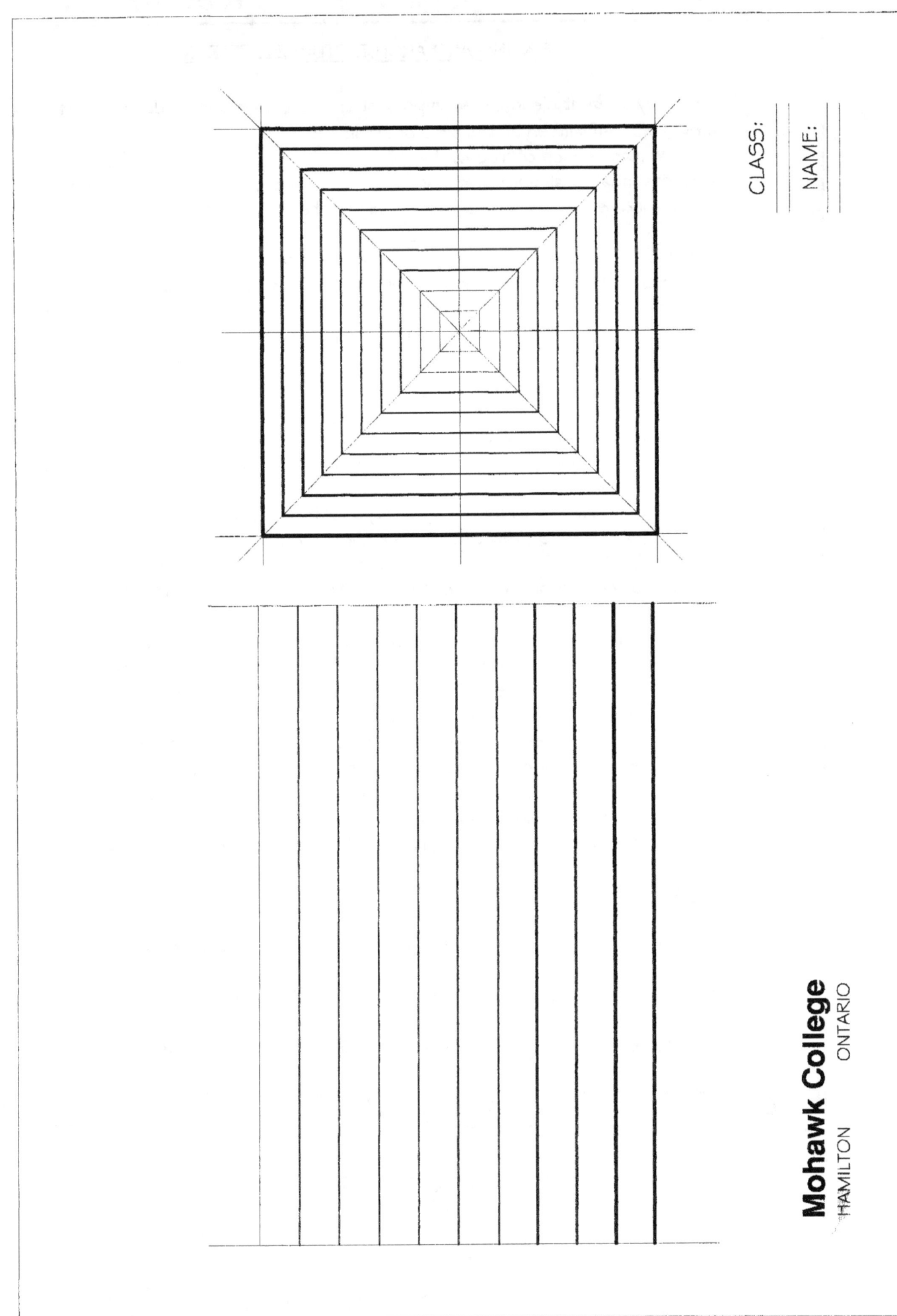

PRINTING AND ARCHITECTURAL LETTERING

Your ability to print clearly is so important that employers may require a lettering sample to be submitted at the time of application!
There are many styles of freehand lettering. The quality and clarity of your printing will be essential to all your drawings. The techniques you use must ensure your work is smudge free, clear and that it follows the industry standards.

Use the following sheets to practice each letter followed by a full printed sheet of technical notes. Do not employ any mannerisms, as they tend to make beginner's notes difficult to read. Let your own printing style develop with time and practice. Concentrate on uniformity, clarity and speed.

Printing Rules:

1. Use capital letters only.
2. Always use *guidelines* when lettering. Draw guidelines very light and never erase them. You can simplify the initial task of measuring locations of guidelines by using a *lettering guide*. With time you will learn how to draw guidelines without aids by estimating line sizes.
3. The form of each letter must appear stable, or "bottom heavy". The lower portions must be drawn slightly larger in area than their upper portions.
4. Compose almost all letters to fit a near-square. Beginners always draw their letters narrower then they should be. Condensed lettering is used only when necessary to fit a very small space. Square form letters are easier to read and look better.
5. Use dark lines for lettering. It is very important that all notes are readable when your drawing is reproduced or reduced in size.
6. Draw letters close together and space words further apart. The spacing is measured by eye so that _areas, not spaces,_ between letters seem equal. Use an imaginary "O" for spacing between words.

Printing Sizes:

1/4" or 5mm	LETTERING FOR TITLES AND DRAWING NUMBERS
1/8" or 3mm	LETTERING FOR HEADINGS AND NOTES
3/32" or 2mm	LETTERING FOR DIMENSIONS AND DIMENSION NOTES

ABCDEFGHIJKLMNOPQRSTUVWXYZ

1234567890

Dorota Goede B.Arch

Mohawk College
HAMILTON   ONTARIO

LETTERING
LETTERING INSTRUCTIONS

CLASS
DATE

DWG

A

B

C

D

E

F

G

H

J

K

L

M

N

O

P

Q

R

S

Mohawk College HAMILTON ONTARIO	PRINTING PRACTICE SHEET	CLASS	DWG
	NAME:	DATE	

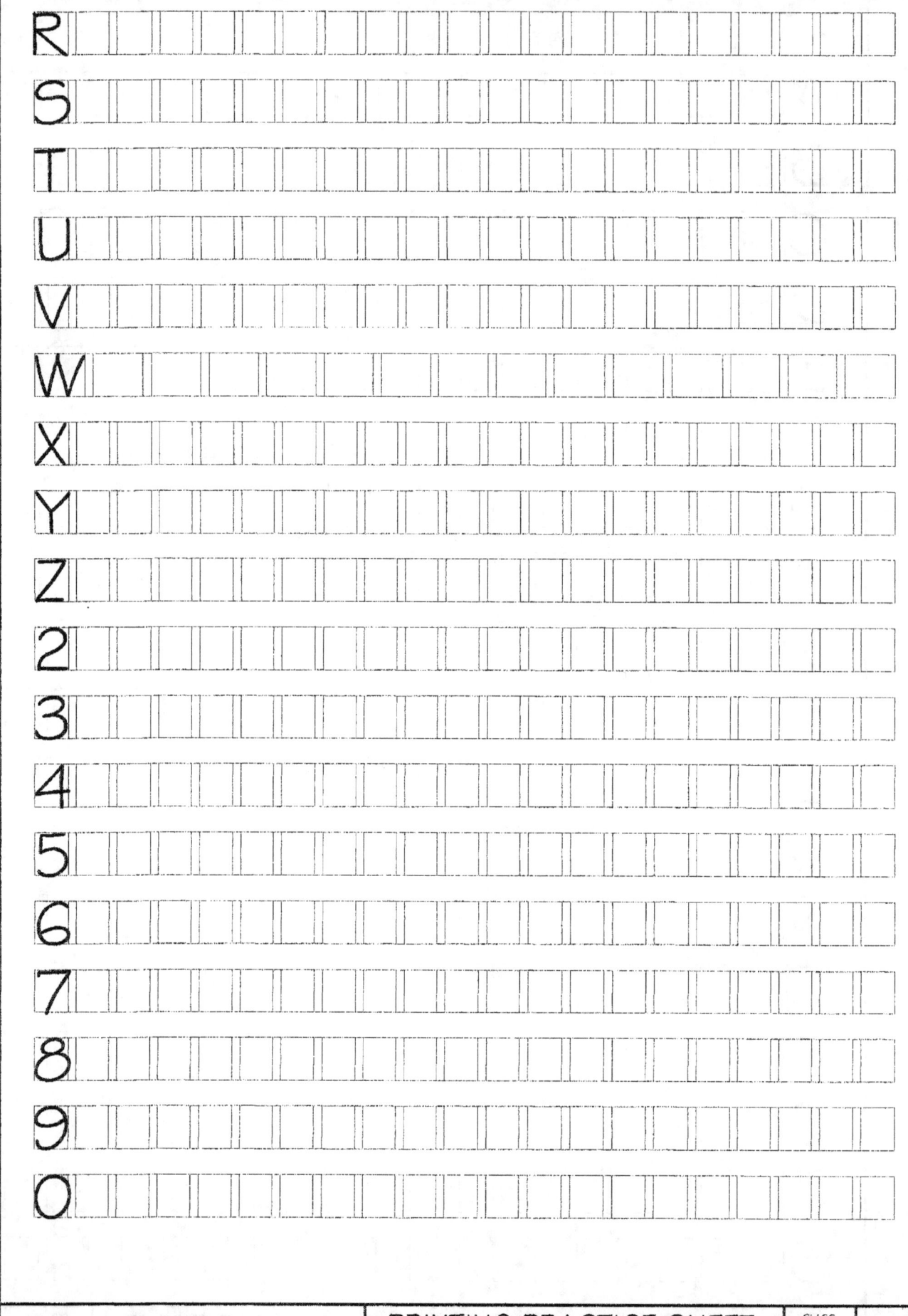

PRINTING PRACTICE SHEET

A
B
C
D
E
F
G
H
I
J
K
L
M
N
O
P
Q
R
S
T
U
V
W
X
Y
Z
2
3
4
5
6
7
8
9
0

Mohawk College
HAMILTON   ONTARIO

PRINTING PRACTICE SHEET

NAME:

CLASS

DATE

DWG

PRINTING PAGE ASSIGNMENT

Use the following two sheets to create a copy of a TYPICAL RESIDENTIAL CONSTRUCTION NOTES page. Learn how to use the lettering guide in the first lab and prepare a page of standard printing lines. Use second lab for copying the printed page attached.

Steps:

- Start with board clean up and set up an A4 size vellum sheet taped square on your board. Find the ergonomically comfortable space near the left board edge and about 6" from the bottom. Set the table level and height if possible. Use capital letters only.
- Prepare 2H sharp lead in your drafting pencil.
- Draw margins allowing for a future trimming to portfolio size 11"X 8.5" size. Draw horizontal and vertical margins accordingly.
- Turn a round part of the lettering guide so that number 4 lines up with a mark on the left non moving side of the guide. See diagram below:

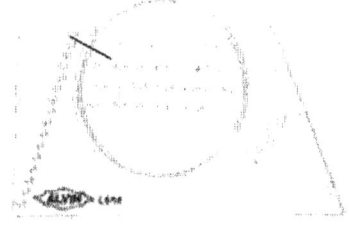

- Place lettering guide over the T-Suare and place a sharp pencil lead vertically in the first looped hole.
- Draw first line gently making sure the lettering guide is guided along the T-square. (press against the T-square with your pencil gently to ensure this!).
- Draw only lines for looped holes and ignore the whole in between. The "looped spaces" will become the printed text and the spaces between loops will become the spaces between text lines. After 3 printing lines you must realign the guide to the last non looped hole. This will ensure the correct pattern. The next required is a space in between lines. Then continue with 3 more looped printing spaces and so on until you fill the entire page with construction guidelines making sure that 3mm and 1.5mm alternate and the pattern is not lost.
- Draw vertical lines to set beginning and end of text, lined up locations for numbers, dots and text offsets.

- Switch pencil lead to a softer H lead and start printing per Lettering instructions on the previous pages. Create a page that could become your printing portfolio.

Dorota Goede B.Arch

Mohawk College HAMILTON ONTARIO | PRINT PAGE ASSIGNMENT | CLASS / DATE | **DWG**

TYPICAL RESIDENTIAL CONSTRUCTION NOTES:

01. ALL LUMBER TO BE SELECT STRUCTURAL GRADE SPRUCE FIR OR PINE
02. ALL STEEL TO BE GRADE 300W PER MIN. CAN3-G40.21 STANDARD.
03. FLOOR AND CEILING JOISTS OVER THE EXISTING HOUSE
    TO BE INSTALLED WITHIN BEAM'S FLANGE TO MINIMIZE CEILING
    EXPOSURE FROM BELOW. DRYWALL CEILING SPACERS MAY BE
    NECESSARY IF OWNER WISHES NO BEAM TO BE VISIBLE.
04. PROVIDE 2"X4" COLLAR TIES 1/3 DOWN FROM THE ROOF PEAK.

FOUNDATIONS:

01. CONCRETE MIX, POURING CURING AND TESTING TO COMPLY WITH CAN3-A438.
02. ALL NEW FOUNDATION WALLS TO BE POURED CONCRETE, 8" WIDTH.
03. PROVIDE FOLLOWING REINFORCEMENT:
    VERTICAL BARS AT ALL CORNERS AND WALL JUNCTIONS SET INTO FOOTINGS.
    STEEL 8"X8" BEARING PLATES AT ALL STEEL POST LOCATIONS TO BE
    WELDED TO POSTS.
04. FIELD VERIFY ALL DIMENSIONS.
05. APPLY CONTINUOUS TWO COATS OF LIQUID TAR OVER NEW FOUNDATION WALLS
    PRIOR TO BACKFILLING. WEEPING TILE TO BE PROVIDED AROUND ENTIRE HOUSE
    TO BE CONNECTED TO CITY STORM SEWER. SUMP C/W PUMP AND DISCHARGE AT
    TERRACE STREET MAY BE INSTALLED IN PLACE OF STORM SEWER HOOK-UP.
    VERIFY WITH OWNER PRIOR TO CONSTRUCTION.
06. MINIMUM OF 24" OF R12 INSULATION TO BE PROVIDED ON FOUNDATION WALLS.
07. ALL SILL MEMBERS LESS THAN 8" ABOVE THE GROUND TO BE OF PRESSURE
    TREATED WOOD.
08. PROVIDE CONC./WOOD GASKETS; 6MIL POLY OR BITUMINOUS SHEETS.
09. NEW FOUNDATION WALLS TO BE LATERALLY JOINED W/EXISTING WALLS

EXTERIOR WALLS:

01. ALL NEW EXTERIOR WALLS TO BE 6" WOOD STUDS ON 16"O/C UNLESS
    OTHERWISE NOTED. RENOVATED WALLS TO BE EXISTING AND NEW 4" STUDS.
    DRAWINGS SHOW ADDITIONAL 2" RIGID INSULATION ADDED ON THE INSIDE
    OF THE EXISTING WALLS. VERIFY WITH OWNER PRIOR TO CONSTRUCTION THE
    EXTENT OF CONSTRUCTION TO BE INCLUDED IN CONTRACT PRICE FOR EXISTING
    STRUCTURE RENOVATIONS.
02. ALL NEW WALLS TO BE FILLED WITH MIN R20 FIBREGLASS INSULATION BATTS.
    ALL EXISTING WALLS TO BE FILLED WITH 4" INSULATION BATTS AFTER NECESSARY
    WALL RECONSTRUCTIONS ARE COMPLETED.

Dorota Goede B.Arch

Mohawk College
HAMILTON   ONTARIO

PRINTING ASSIGNMENT

CLASS

NAME:

DATE

DWG

SAUNA MATERIAL KIT CONSTRUCTION NOTES:

1. **FRAMING**
   USE 2" X 4" CONSTRUCTION GRADE SPRUCE STUDS TO FRAME THE WALLS AND CEILING. IF THE CEILING SPAN IS OVER 10', USE 2" X 6" STUDS. UNLESS OTHERWISE SPECIFIED, FOR SAUNAS UP TO 8' X 8', THE WALLS ARE LINED HORIZONTALLY AND FRAMED VERTICALLY. FRAME THE WALLS AT 16" ON CENTER. FRAME OR STRAP THE CEILING TO AN INSIDE HEIGHT OF 82-1/2". FRAME THE CEILING SO THAT THE CEDAR WILL RUN THE SHORTEST DIRECTION. PROVIDE EXTRA BLOCKING FOR HEATER.

2. **WIRING**
   THE THERMOSTAT SHOULD BE INSTALLED ON THE WALL OUTSIDE THE SAUNA AND THE SENSING BULB SHOULD BE PLACED AT THE CEILING ABOVE THE HEATER LOCATION (SEE ELECTRICAL WIRING DIAGRAMS WITH HEATER AND CONTROL). HOOK-UP OF THE HEATER AND LIGHT CAN BE DONE AFTER THE SAUNA IS COMPLETED.

3. **INSULATION**
   THE SPACE BETWEEN THE STUDS IN THE WALLS AND CEILING IS FILLED WITH 3-1/2", R-12 FIBERGLASS INSULATION.

4. **VAPOUR BARRIER**
   FOIL VAPOUR BARRIER OVER THE INSIDE EDGE FACE OF STUDS OVER THE INSULATION.

5. **STRAPPING**
   IF LINING VERTICALLY, STRAP (1" X 2" OR 1" X 3") OVER THE FOIL TO CHANGE DIRECTION. CAN BE USED TO CREATE AIR SPACE, BUT IT IS NOT NECESSARY.

6. **LINING**
   USE THE SPECIFIED LENGTHS OF TONGUE AND GROOVE CEDAR. TRIM BOARDS. NAIL THROUGH THE TONGUE ONLY (BLIND NAILING). USE THE 1-1/2" NAILS SUPPLIED. INSTALL THE TONGUE AND GROOVE CEDAR TO THE CEILING FIRST. START LINING THE WALLS AT THE FLOOR AND WORK YOUR WAY UP. THE FIRST CEDAR BOARD SHOULD BE LEFT ABOUT 3/4" OFF THE FLOOR.

7. **FASTENERS**
   FASTENERS SUPPLIED ARE GALVANIZED NAILS AND DECK KING TREATED SCREWS. ALL FASTENING IS TO BE BLIND NAILED.

8. **BENCHES**
   THERE ARE CLEAR CEDAR 2" X 4" BOARDS FOR THE BENCH CONSTRUCTION. SELECT THE BOARDS TO MAKE SURE THE BETTER FACE OF THE BOARD COMES OUT ON THE SITTING SURFACE. WHEN BUILDING THE BENCHES, FASTEN FROM THE BOTTOM. USE THE 2-1/2" TREATED SCREWS SUPPLIED. IF THE SAUNA IS 5' OR MORE DEEP, THE TOP BENCH IS 20" DEEP AND THE BOTTOM BENCH IS 16" DEEP.

9. **DOOR**
   SOLID CEDAR DOOR WITH 12" X 24" INSULATED DOUBLE PANE WINDOW OR CEDAR FRAMED FULL LENGTH GLASS DOOR. DOOR OPENS OUT. PRE-HUNG ON 3/4" X 4-9/16" MAHOGANY JAMB. DOOR SIZE IS 24" X 74". ROUGH OPENING IS 26"X76".

Mohawk College HAMILTON ONTARIO	FREEHAND PRINTING	CLASS	DWG
	NAME:	DATE	

BUILDING PERIMETER DRAFTING AND CALCULATION ASSIGNMENT

ASSIGNMENT 1:

Draw three building plans as shown on the following page. View this page in "landscape" orientation! Use construction lines to outline the A2 sheet layout in scale 1:100. Design vertical title block, 50mm wide on the right sheet side. Use standard title block and margins as outlined in the manual's Appendix. Space titles centrally within designated boxes and use 5-7mm high texts for titles. The corner box will always be located at the right bottom sheet corner and it must contain scale, due date, your class, and your name information.

The objective of this assignment is to further practice line drafting of minimum three varied line weights, to design tight sheet layout, and to introduce 45deg hatching using very light lines at 3mm apart. Use your 45/45 triangle along T-square and eye set the spacing. Highlight the concrete floor; hatch lines span from inside faces of the interior walls.

Dimensioning techniques will also be introduced. Study dimensioning sheet available in the Appendix. Please note that all dimensioning must be done using arrows. All dimension notes must be outlined with two guidelines and texts must be 3mm high.

All walls are 300mm wide and all footings are 900mm.

Refer to the Appendix pages for information about required Department standards for sheet sizes and Titleblock borders.

ASSIGNMENT 2: CALCULATE THE PERIMETERS OF THE OUTSIDE FOR ALL THREE PLANS.

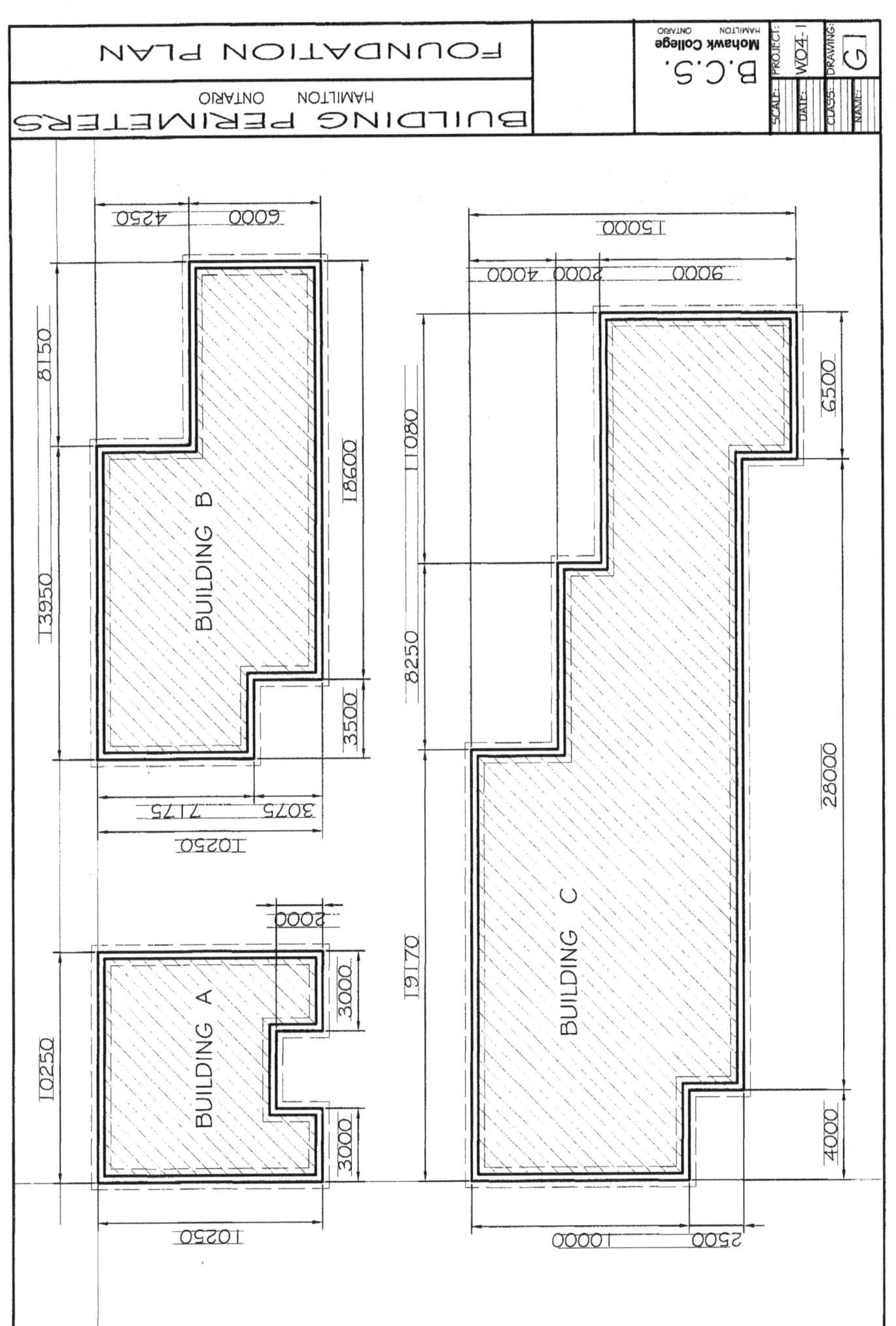

BASIC GEOMETRIC CONSTRUCTIONS

The following page explains the most common techniques used to draw some of the basic geometric constructions. They are so elemental all drafters use them on every drawing in some form. Make sure you learn these techniques, as your employer will expect you to be precise in your geometry.

Your drafting equipment includes two drafting triangles: 30/60/90 and 45/45/90. The six techniques on the next page explain how to use the two triangles in order to draw parallel and perpendicular lines, lines tangent to arc, equilateral triangles and squares. The dashed line triangle represents the triangle moved to the new position.

Practice all 6 techniques on the attached three work sheets. Remove the sheets from your book, complete all titleblocks, draw, staple when finished and submit to your Instructor for evaluation.

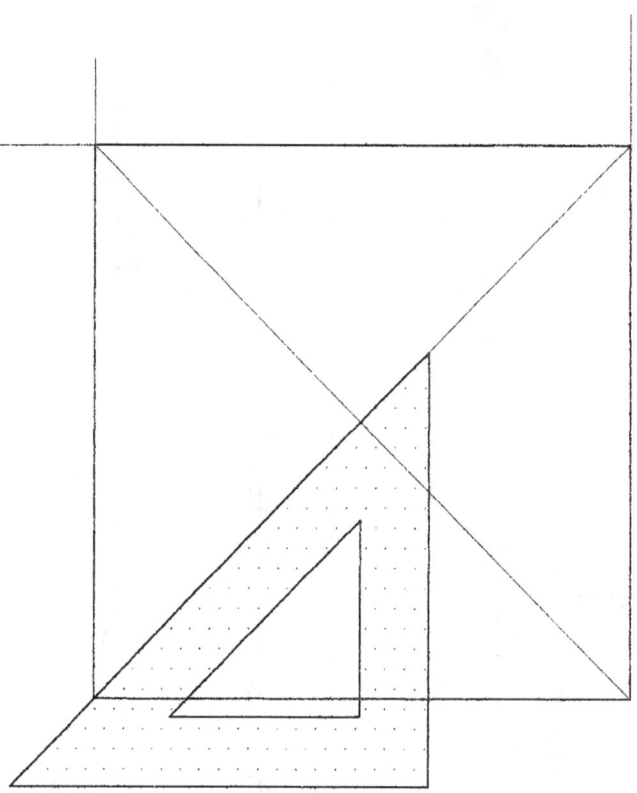

Dorota Goede B.Arch

Mohawk College
HAMILTON   ONTARIO

GEOMETRIC CONSTRUCTIONS
INSTRUCTIONS

CLASS
DATE
DWG

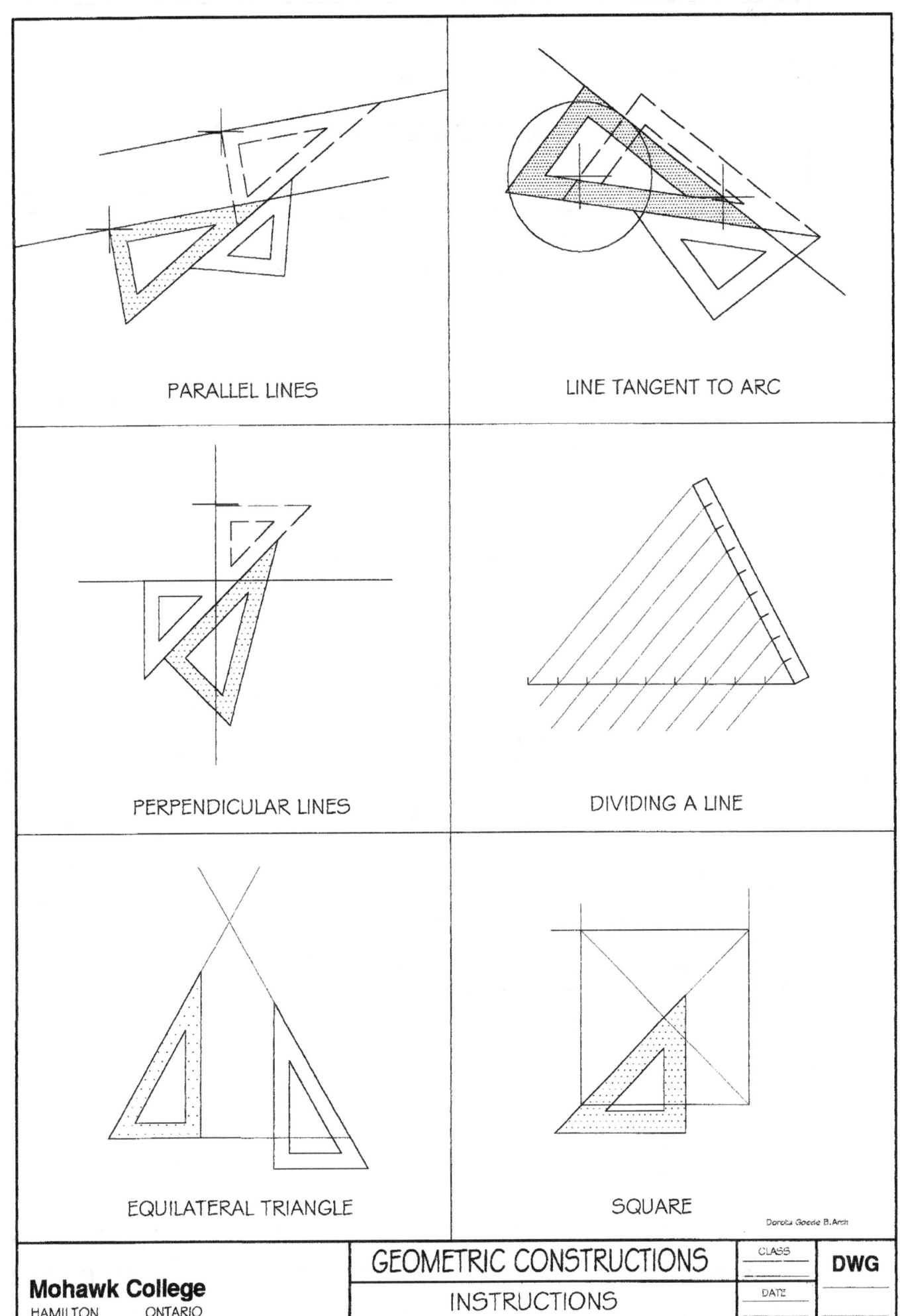

ANGLE CONSTRUCTIONS ASSIGNMENT

The drawing below explains the most common techniques used to draw lines with a specific angle from the horizontal base. Use the two drafting triangles: 30/60/90 and 45/45/90.

Practice all constructions shown on the following page.

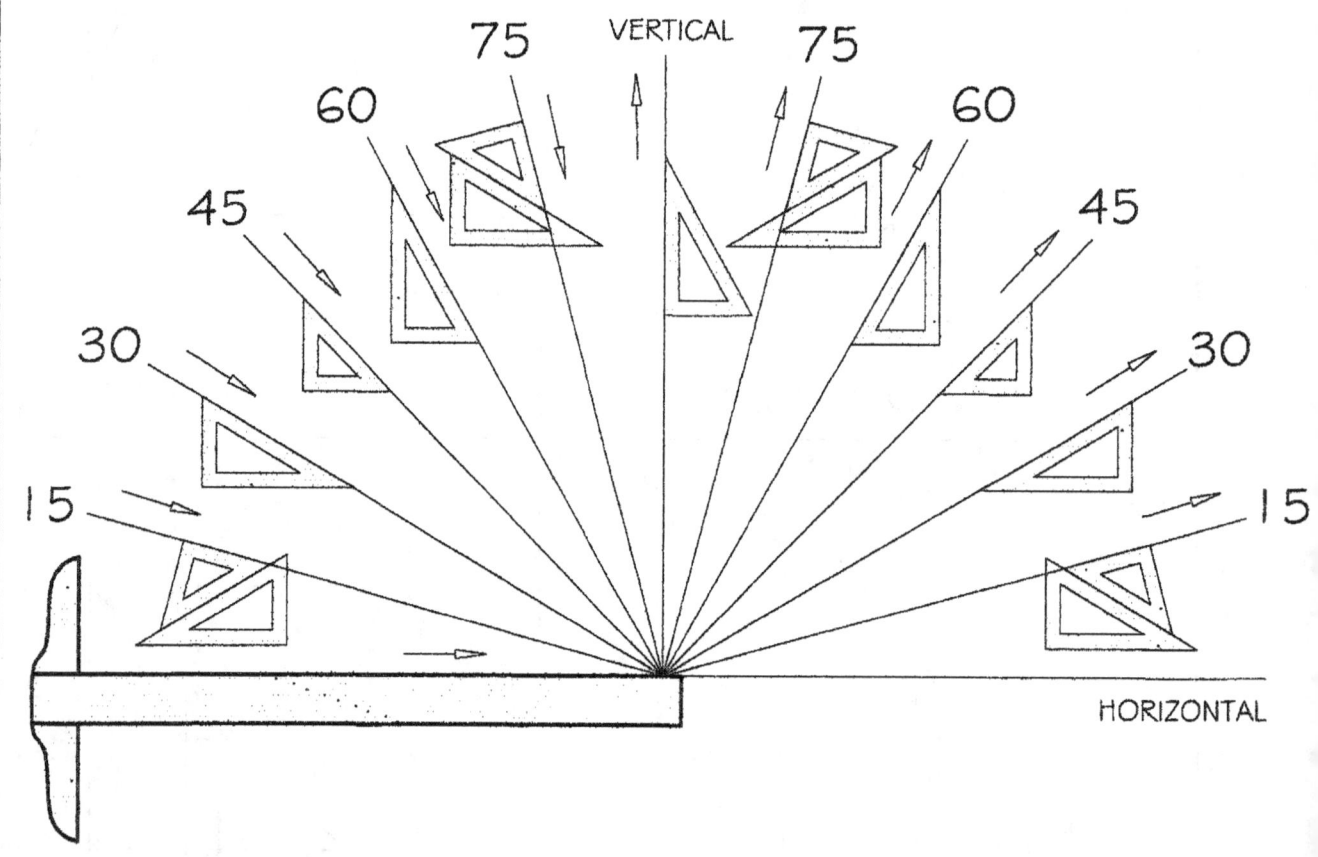

Mohawk College
HAMILTON   ONTARIO

INSTRUCTIONS

ANGLE CONSTRUCTIONS

NAME:

CLASS

DATE

DWG

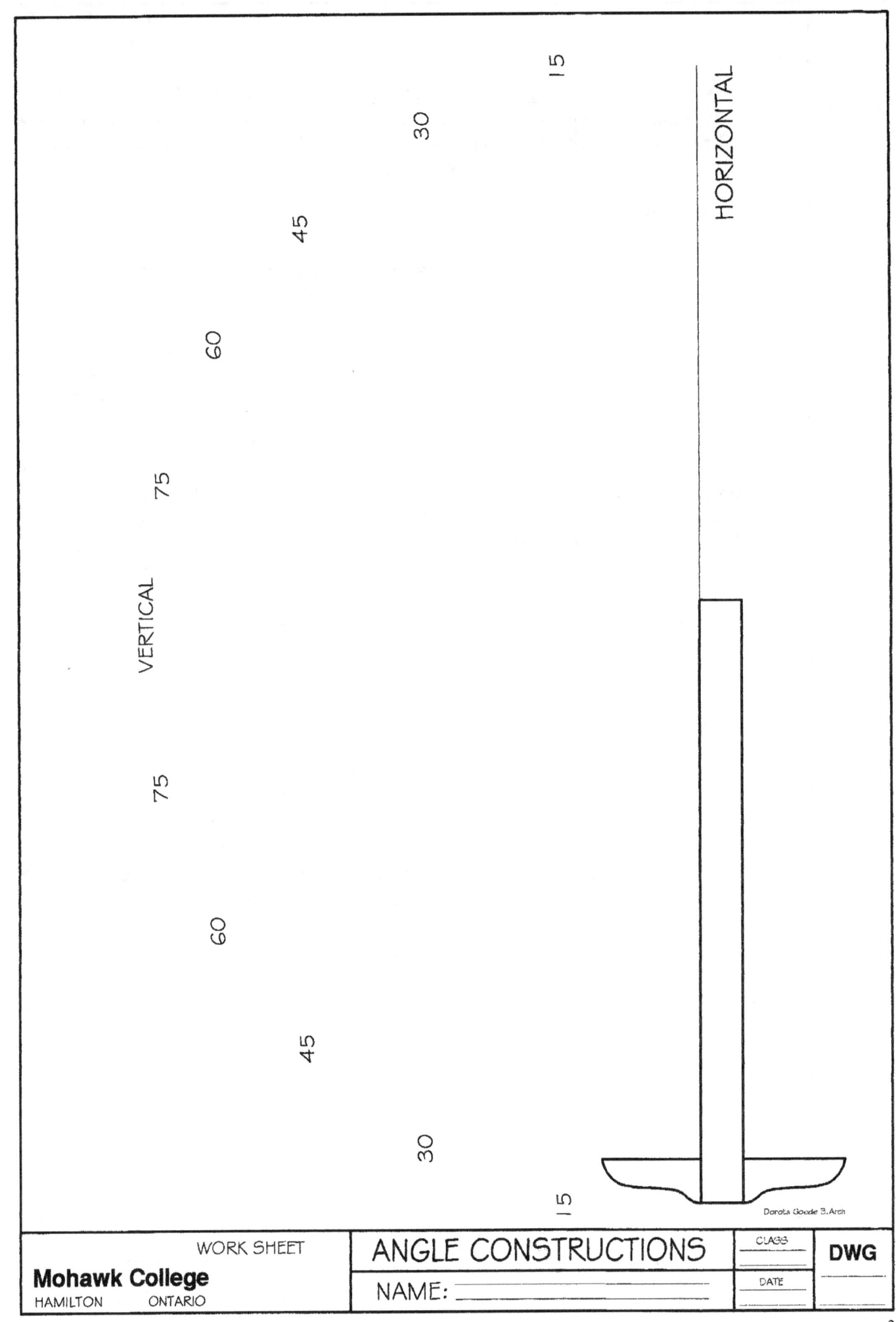

WOLMANIZED WOOD DECK DRAFTING PROJECT

The deck, as shown on this page, can be proposed as a small balcony, or deck, off the sliding door to the outside. Given is a plan, in ½"= 1'0" scale, and a layout complete with the required dimensions.

ASSIGNMENT #1:

Draw Plan and design the typical Elevation for the deck shown. Propose ELEVATION A-A as marked on the plan. Use A3 size vellum vertically and predetermine your layout for drawings required. Draw plan layout at 1"=1'-0" scale, as shown. Double-check all dimensions. Allow room for the elevation drawing. Drawing layout and the accuracy of your work is always drafter's responsibility. Use light construction lines to project all information required for the Elevations. In this exercise the given information on the plan should be projected down via light construction lines!

Have fun designing the railing. Main posts are 4"X4". The standard design is 2"X2" rail posts with 4" spaces in between. These are usually mounted at the bottom support member parallel to the floor and mounted 4" above it. The typical balcony rail, 2' above ground must be 36" high per the requirements of the Ontario Building Code. The floor level is 6ft above the ground and the required pressure treated wood support beams will be 2"X10". Floor construction will be a typical 2"X8" joists @ 16" centers.

Draw standard department Titleblock per Perimeter assignment.

ASSIGNMENT #2:

Determine the material required for construction purposes. Obtain up-to-date prices from your local lumber store. Draw table on your drawing and allow space for your CONSTRUCTION NOTES. Specify lumber, joints, hardware, finishes etc. Design pier foundations and concrete to wood connections. Library and local building centers will assist you researching standard construction methods.

ITEM	MEMBER SIZE	LENGTH REQ'D	STANDARD LENGTH	UNIT PRICE	TOTAL PRICE	LUMBER TOTAL
					Total:	$

Mohawk College
HAMILTON ONTARIO

DECK DRAFTING PROJECT
INSTRUCTIONS

CLASS
DATE
DWG

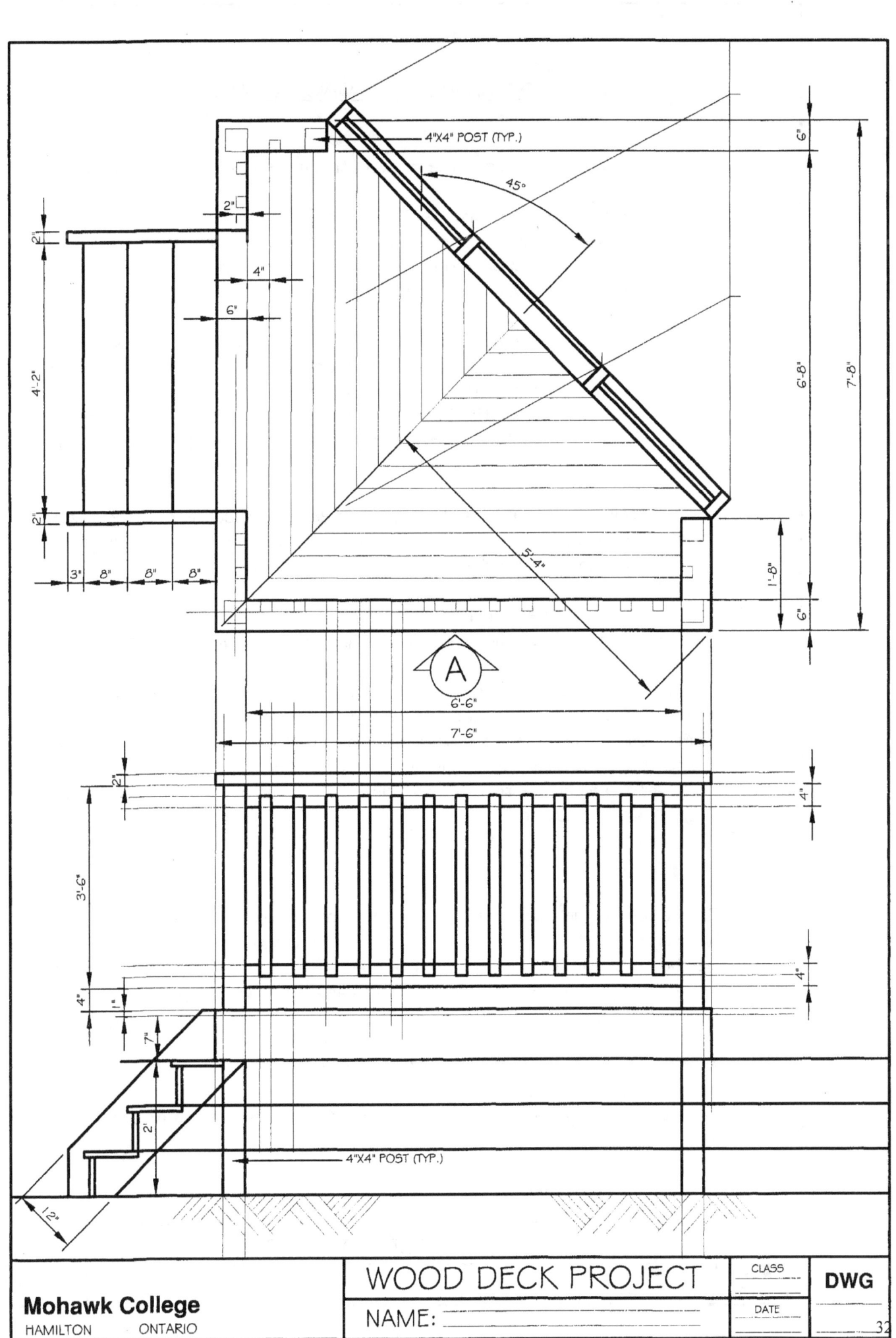

FREEHAND LINE DRAWING TECHNIQUES

There are number of freehand line drawing techniques available to help you:

1. A freehand line is not perfectly straight and accurate. The unevenness is what makes such lines more interesting and appealing. The line's weight, direction and proportions representing distances between objects are very important. <u>Draw freehand lines long and continuous.</u>

2. Use the edge of your drawing board to guide your smallest finger in order to draw long, straight horizontal and vertical lines, especially borderlines. Use the hand holding the pencil and stiffen it as you draw and move along the board edge. Stretch your hand and fingers to draw parallel lines. Turn your baseboard around 90deg to draw horizontal lines using the same technique.

3. All angled, long lines can be checked for correctness by looking at them at eye level. To do this lift the board and turn accordingly. Look at the line from the beginning of the and towards the end.

4. Use soft pencil lead only. Turn pencil around so the end does not flatten. The pencil should be slightly rounded and not needle-sharp.

Observe, compare and measure by comparing the object proportions. This will immediately distinguish an excellent work from a mediocre one. The overall proportions of the object are especially important as they set the proportions of all smaller object elements. Many building components are so common that it is worthwhile to memorize their proportions: 2"X4" lumber, 8"X16" concrete block, 2 2/3"X4"X8" brick etc.

Assignment 1:

Draw freehand the front and back of the church-like form shown on the following two pages. Use 2 of sheets 11"X17" size bond paper. Maximize the drawing size but allow for fair size margins all around. Start by drawing lightly the cube with 4X4 divisions. Establish crucial points and recreate the geometry.
You may add more detail to imply doors, windows or steps. If shading use only light straight lines applied in varied directions.
Your drawing should differ as it will be freehand. All lines should be extended beyond desired points.
Do not preoccupy yourself with perfection. Your freehand drawing should be fluid and layered. Have fun with it.

Dorota Gorde B.Arch

Mohawk College
HAMILTON   ONTARIO

FREEHAND SKETCHING
ON SITE DRAWING EXERCISES

CLASS
DATE
DWG

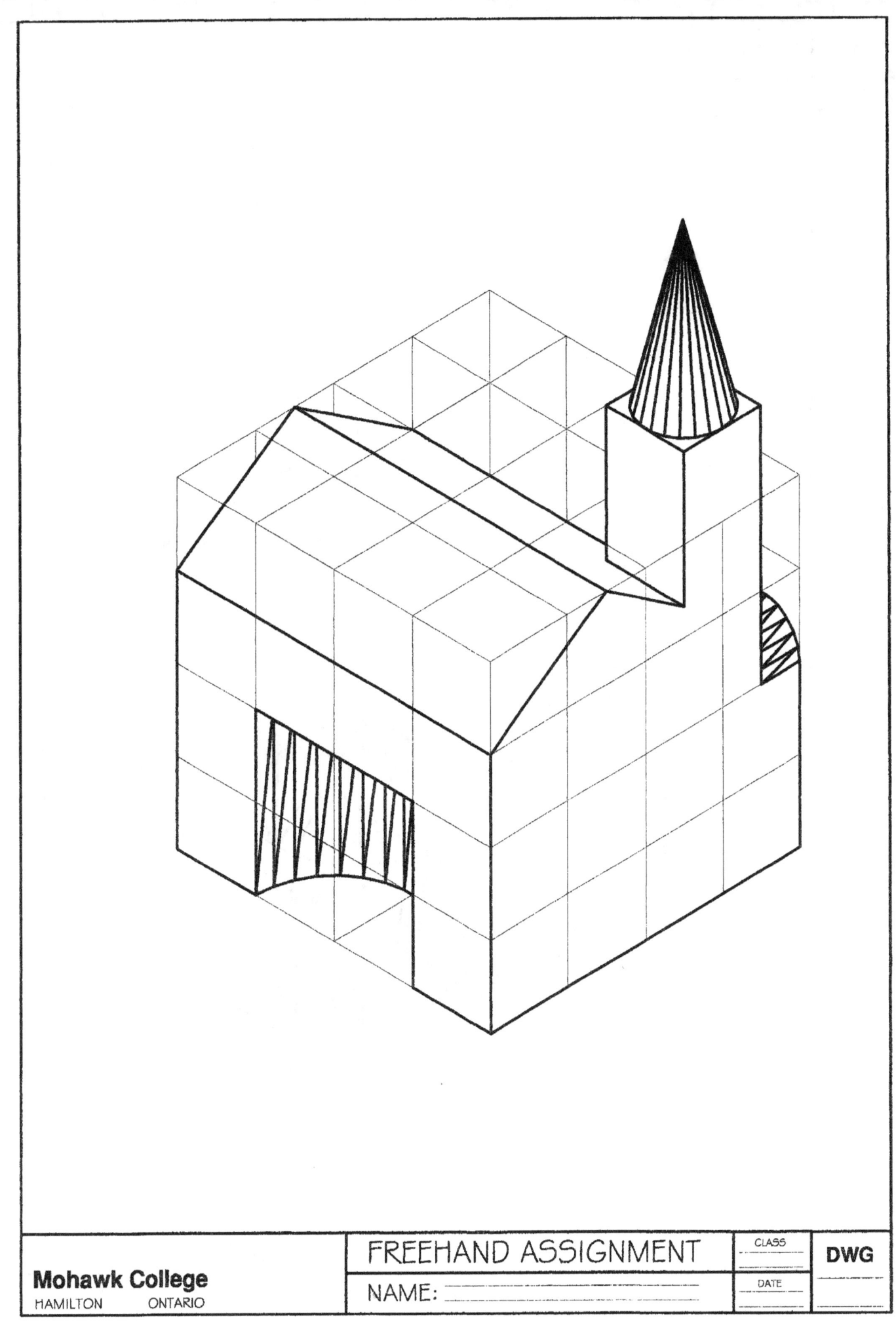

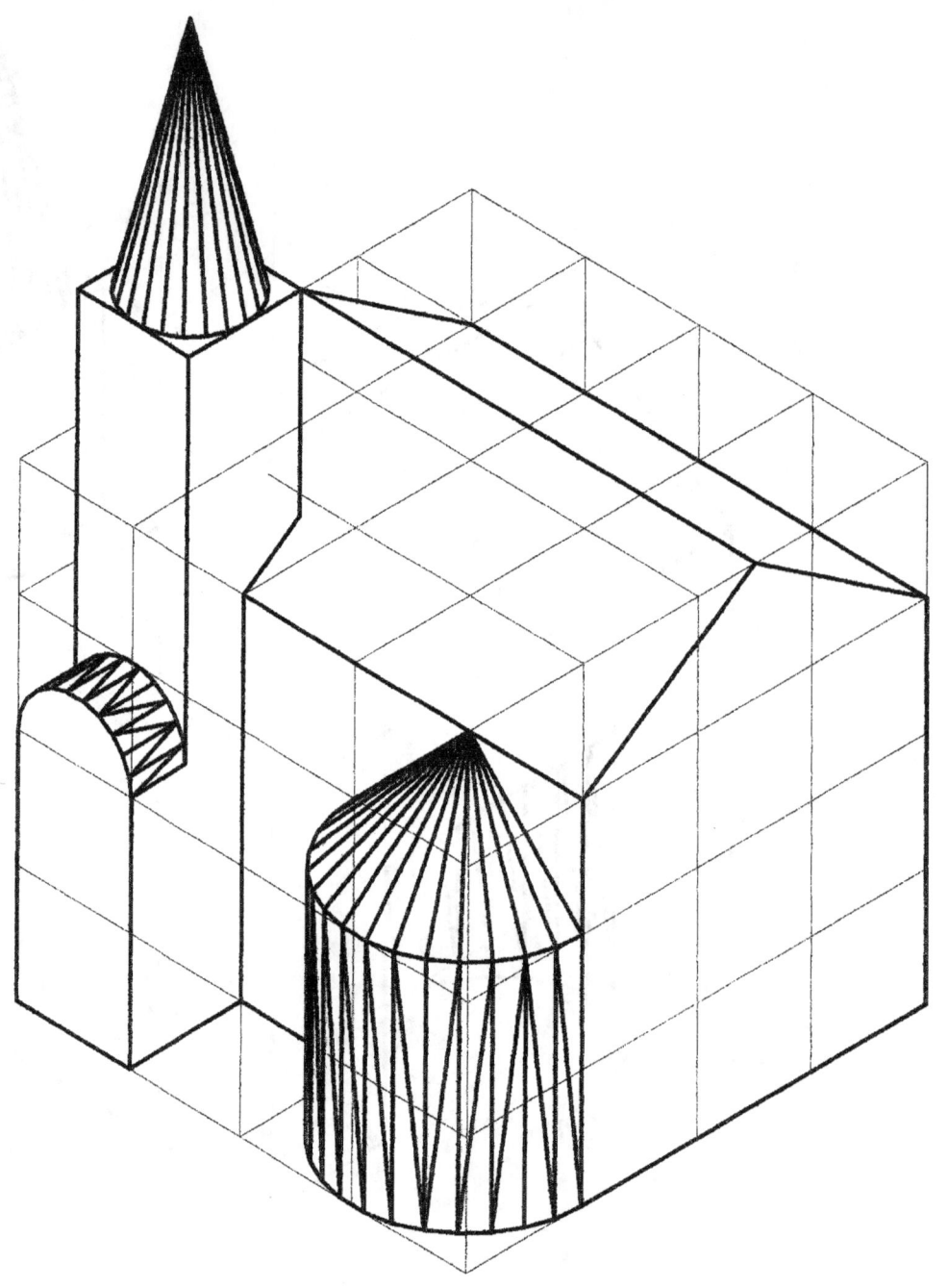

FREEHAND SKETCHING ; LINE DRAWING INSTRUCTIONS

You do not have to be born artistic in order to learn how to draw! You can become very proficient in freehand sketching and drawing with practice and proper guidance.

As a member of any design team, you must be able to convey your ideas. You must be able to clarify your points during discussions and sketch quick revisions and solutions to problems. Artistic representations of your designs can explain the final project to developers, investors, loan officers etc. Construction detail solutions are often sketched by designers and then drafted to scale. Freehand sketches always accompany photographs when existing structures are measured and drafted.

Sketching and freehand drawing are integral components of your professional training.

Assignment 2:
Copy line exercises on the next page in larger scale on following blank sheets provided. Use light construction lines to locate intersections. A dotted grid will help you with proportions.

Assignment 3:
The following freehand drawing should be drawn live from a set composition or some existing scenario. Your instructor will assign the topic of this exercise. Weather allowing you may be asked to draw some existing detail or a simple structure on Mohawk's property. The following are some suggestions:

1. Winding stair model in the A220 model shop; Plans, Elevations and 3D freehand sketch.
2. Elevation of the J wing building.
3. Plan and section of the roof truss supporting the Gym roof.
4. Plan, elevations and 3D representation of a column base detail complete with anchor bolts.
5. Bus shelter three dimensional sketch.
6. Plan of the front entrance parking lot c/w ticket booth and landscaping.

Prepare some kind of smooth surface base board or Bainbridge card. There are some boards you may borrow from the model shop A220. Please do not damage them and put them back so other students can reuse them.

Dorota Goede B.Arch

Mohawk College
HAMILTON   ONTARIO

FREEHAND SKETCHING
ON SITE DRAWING EXERCISES

CLASS
DATE

DWG

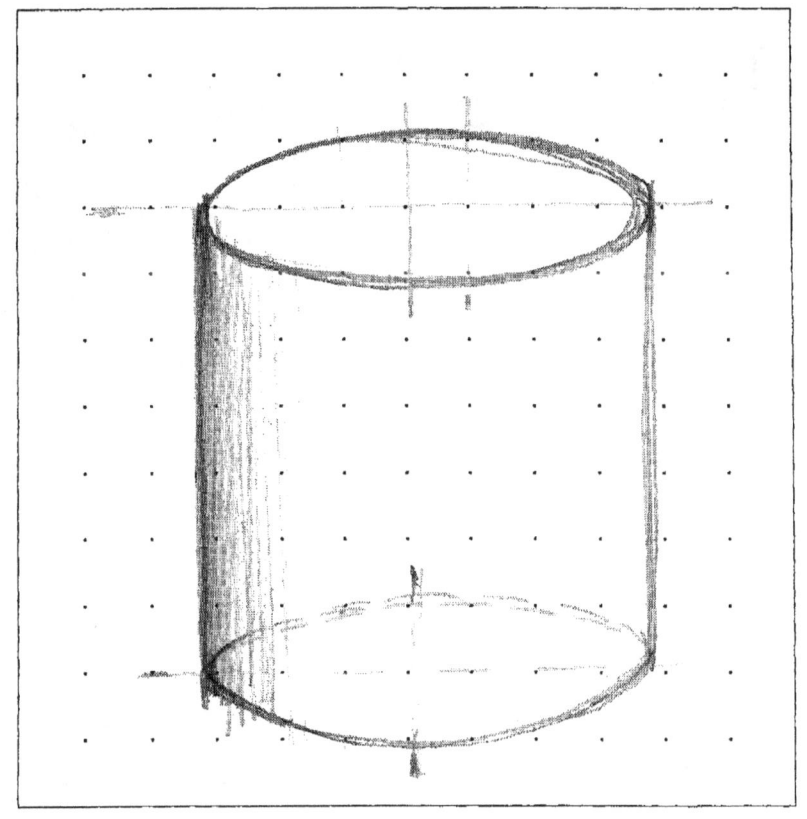

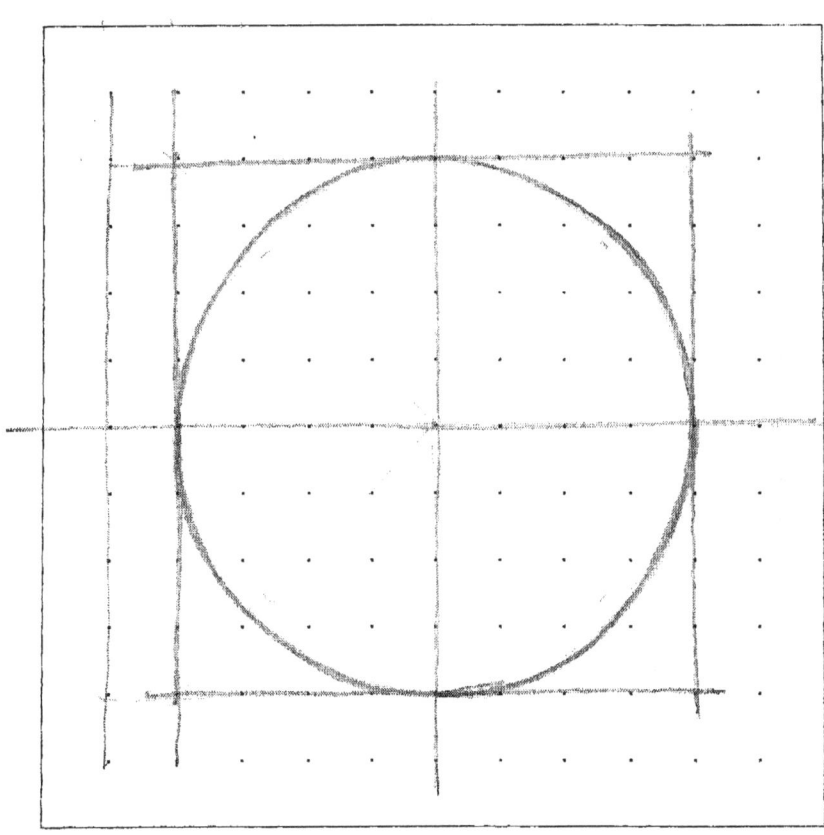

FREEHAND SKETCHING	CLASS	DWG
NAME:	DATE	

Mohawk College
HAMILTON   ONTARIO

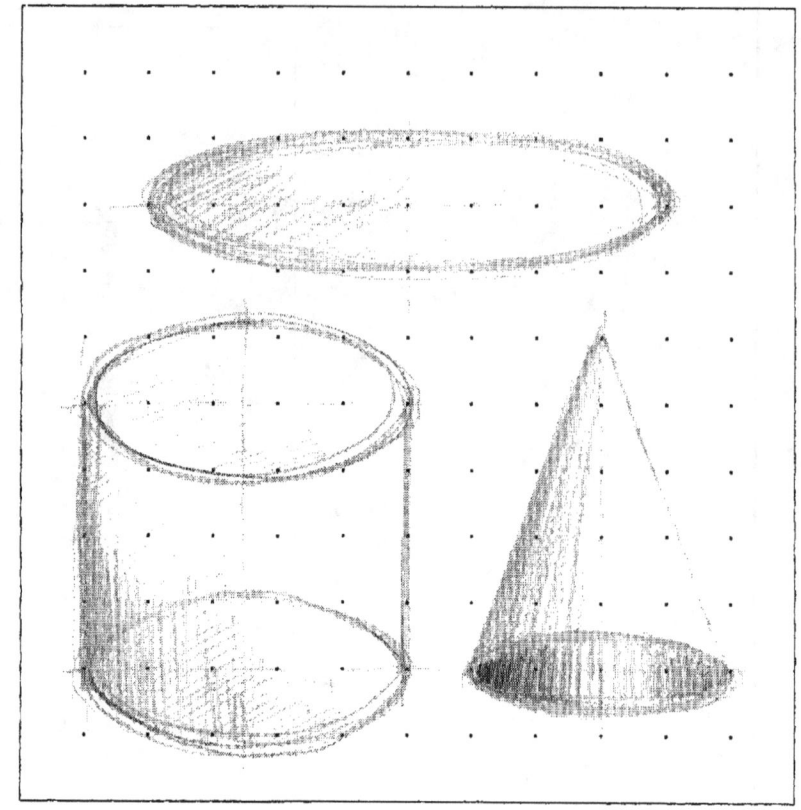

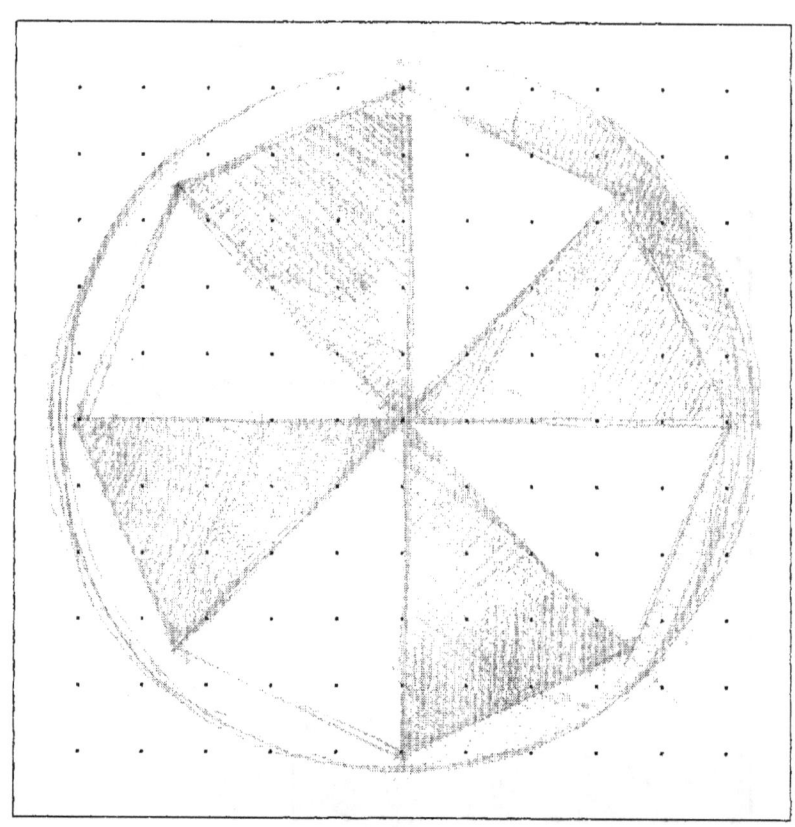

Mohawk College
HAMILTON   ONTARIO

FREEHAND SKETCHING

NAME:

CLASS

DATE

DWG

FREEHAND SKETCHING | Mohawk College — Hamilton, Ontario

NAME: _____ CLASS _____ DATE _____ DWG

FREEHAND SKETCHING

NAME:

CLASS

DATE

DWG

Mohawk College
HAMILTON   ONTARIO

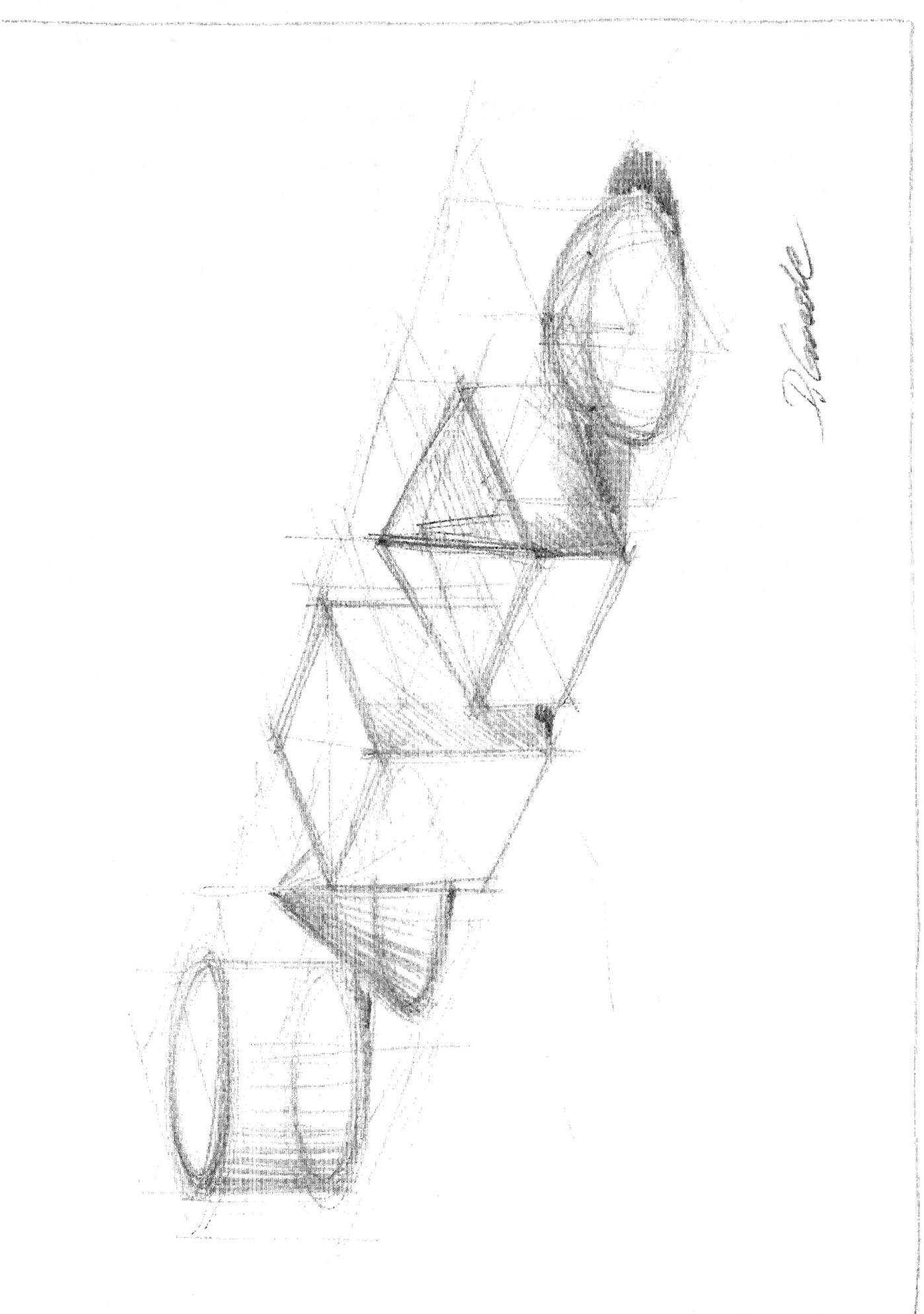

ADVANCED FREEHAND DRAWING ASSIGNMENT

Use the following Isometric steel construction detail and create your own 3d representation of the steel roof elements like girders, beams and columns.
Use 11"X 17" bond sheet and HB soft pencil lead. Draw light construction lines first. All parallel lines must be parallel on your drawing as well.
Use 30 deg angle for both horizontal all horizontal lines on the drawing.

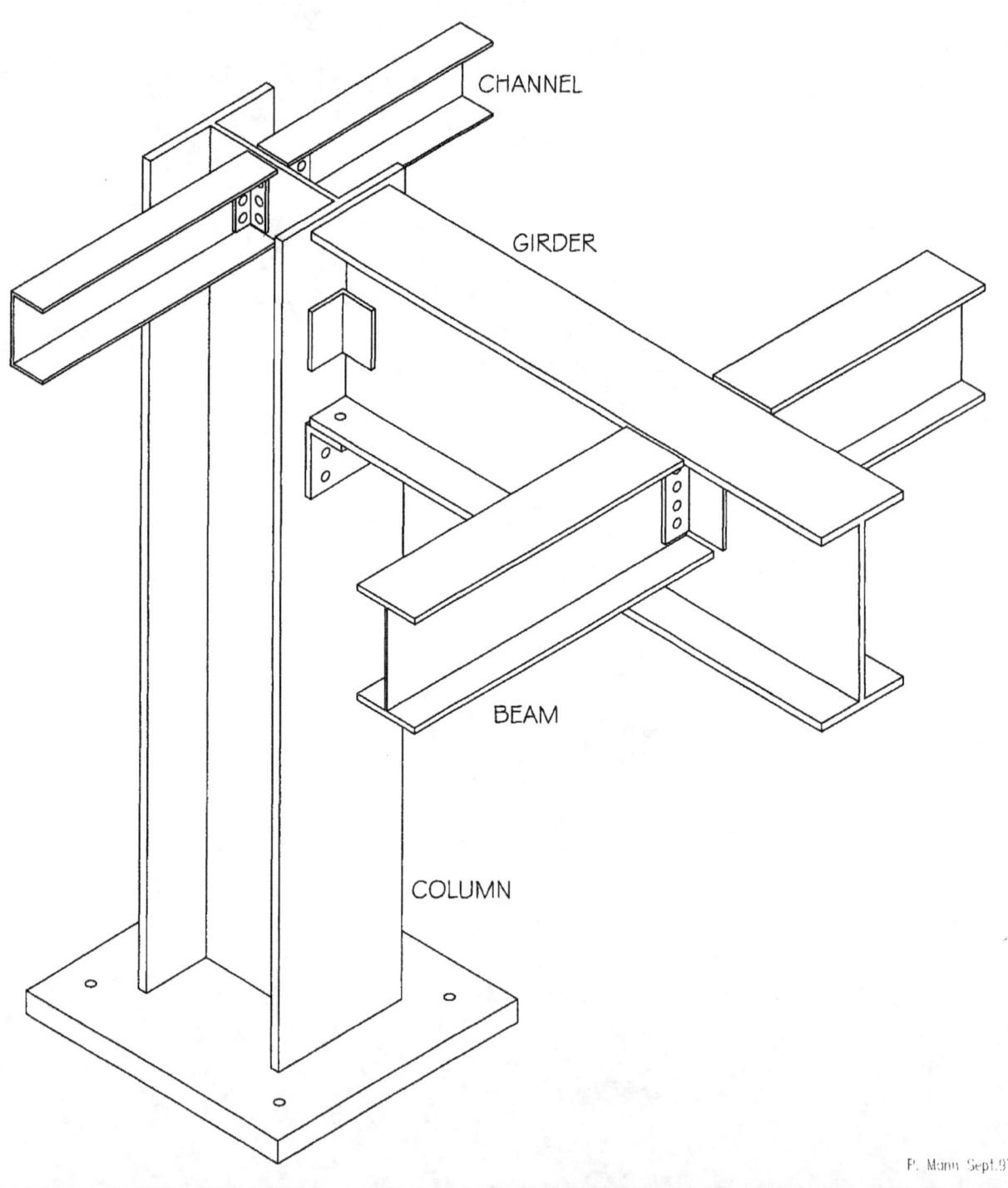

STEEL FRAME ROOF PLAN

Mohawk College
HAMILTON   ONTARIO

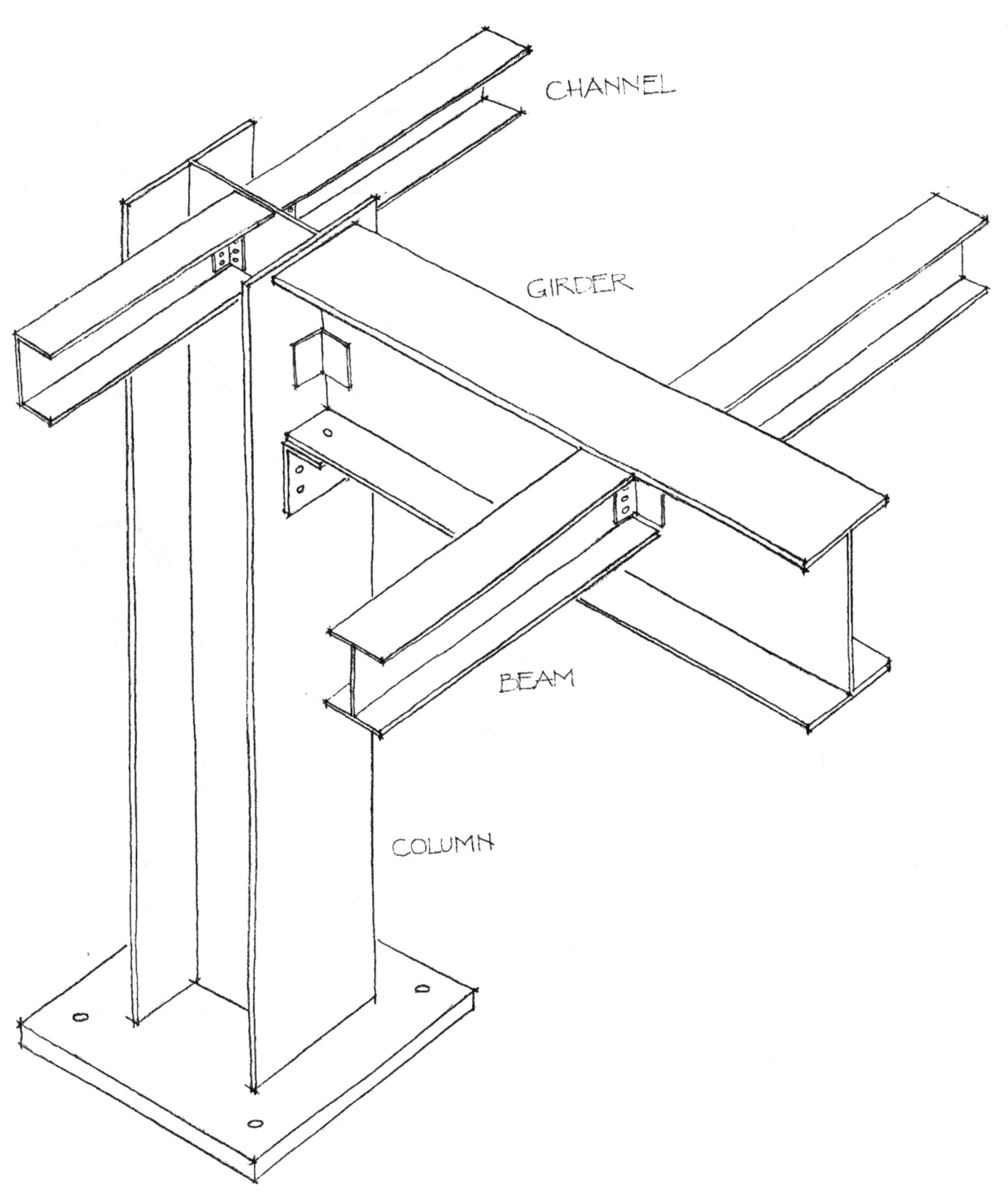

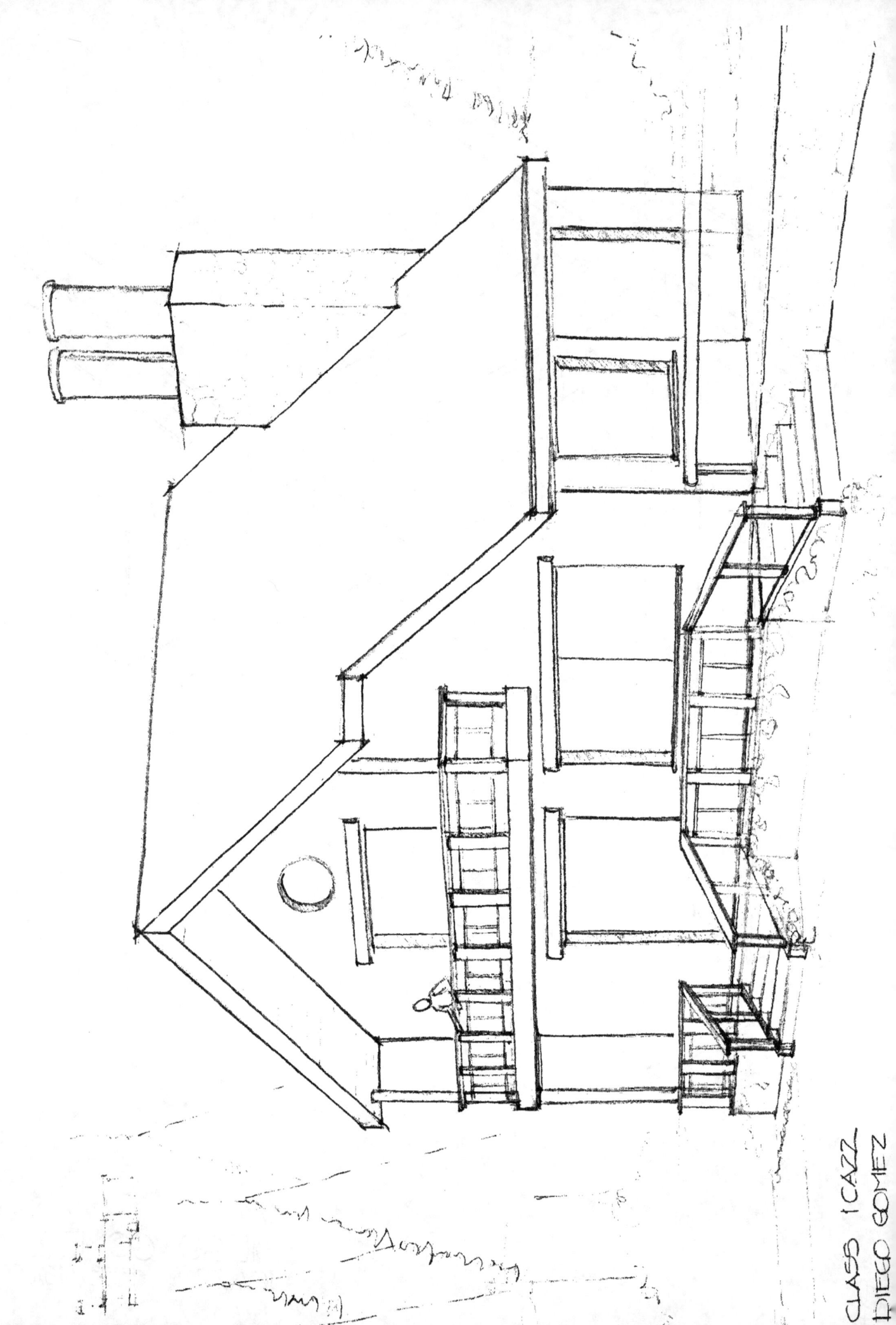

ORTHOGRAPHIC PROJECTION ASSIGNMENTS

The drawing below represents an isometric view of a cube with parts of it "cut out". Noted are elevation views and a North arrow. The next page represents elevation views drafted in 2D. The top part shows the first stage, only closest seen, surface edges. The bottom drawing shows the final solution complete with all hidden lines.

Assignment 1:

Practice 3D/2D visualization by completing the attached assignment pages.

Two isometric views are shown for each one of the six compositions. Observe and study the isometric projections and draw the Top, Bottom, South, East, North and West Elevation views. The allocated spaces for each composition are numbered in circles. The exercise pages are marked with dots to aid the correct proportions. Your Instructor will request freehand or drafted sheets to be handed in.

Remove the work sheets from the manual book. Draw all lines representing the surfaces first seen, darkest. Show invisible edges with much lighter and dashed line. All surface edges must be shown. Submit your work for evaluation.

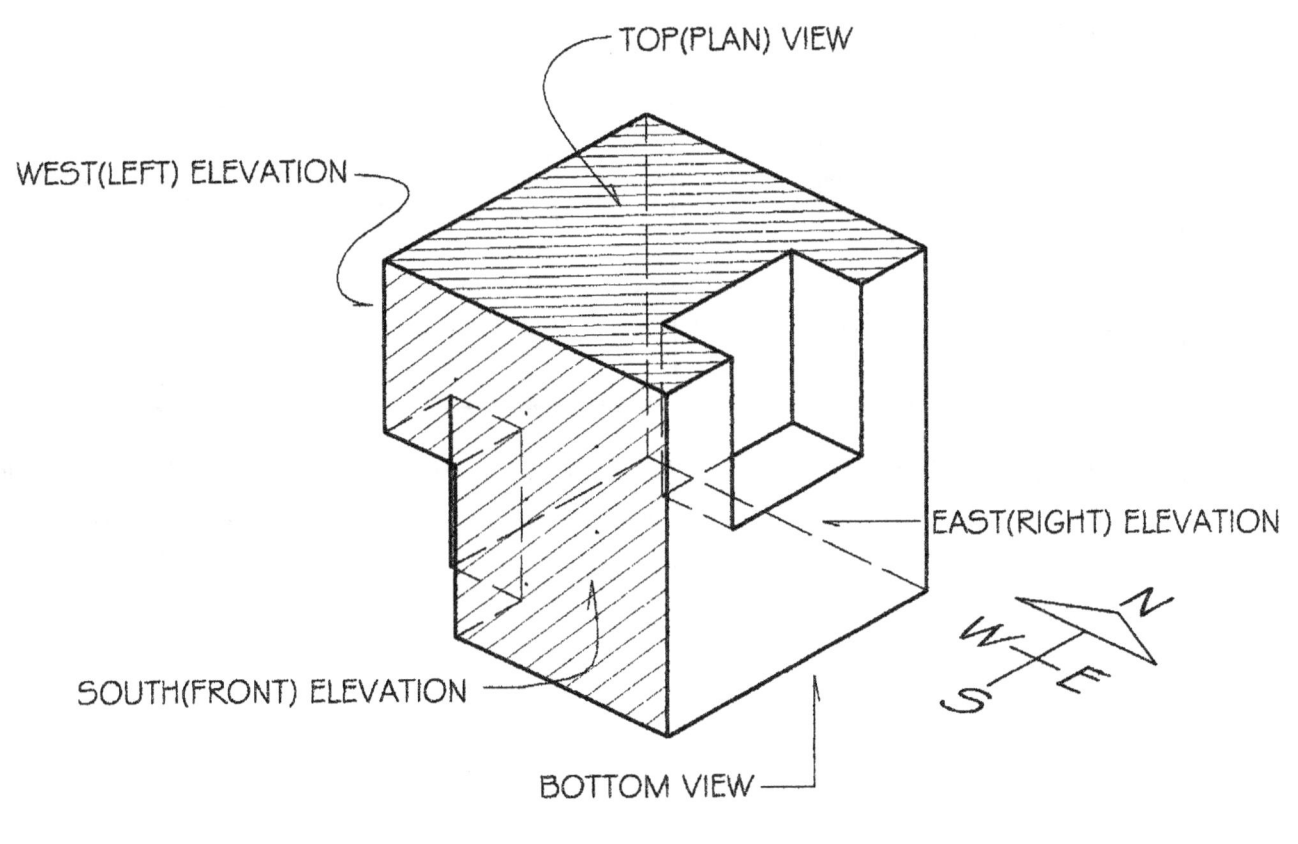

	ORTHOGRAPHIC PROJECTIONS	CLASS	DWG
Mohawk College HAMILTON ONTARIO	INSTRUCTIONS	DATE	

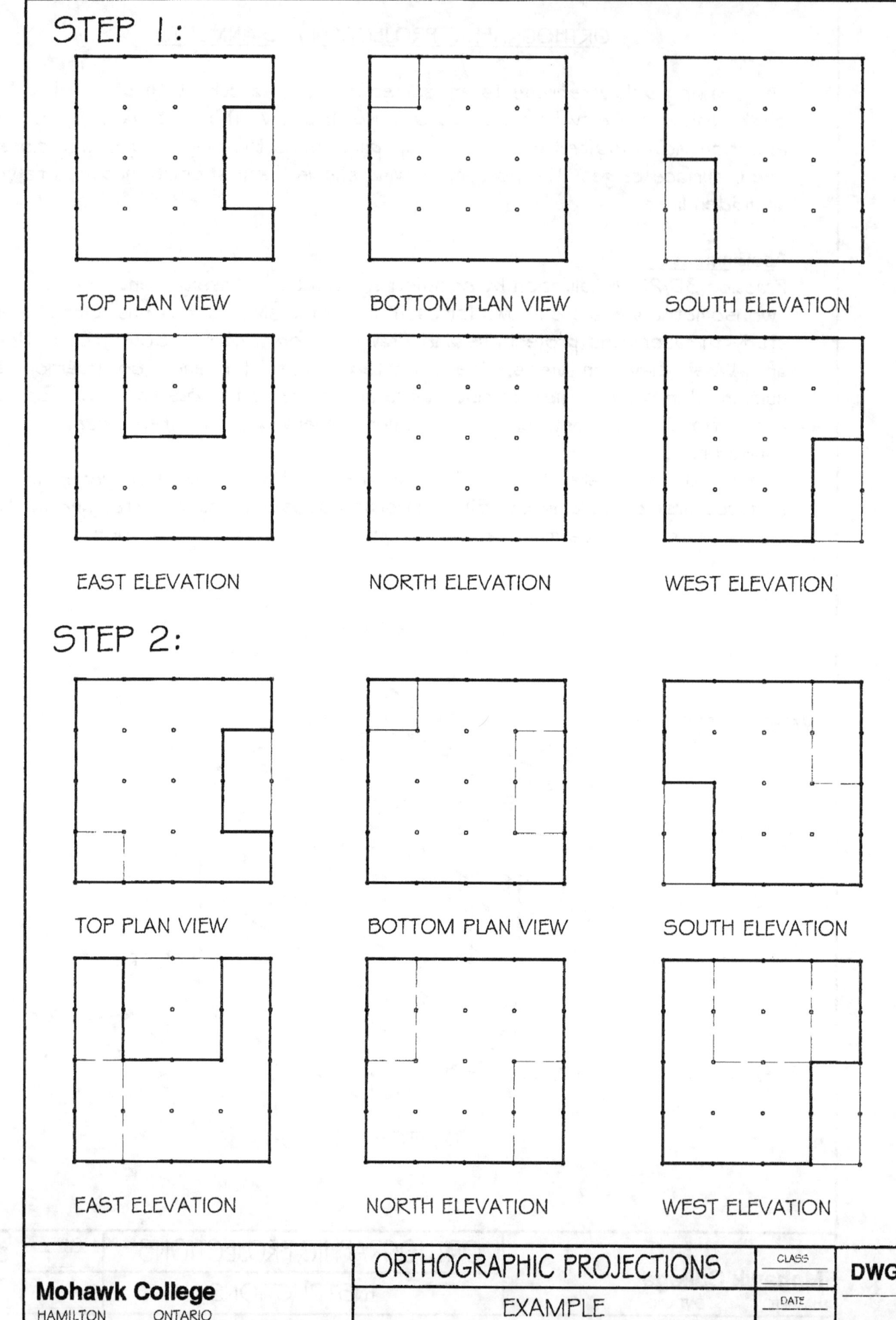

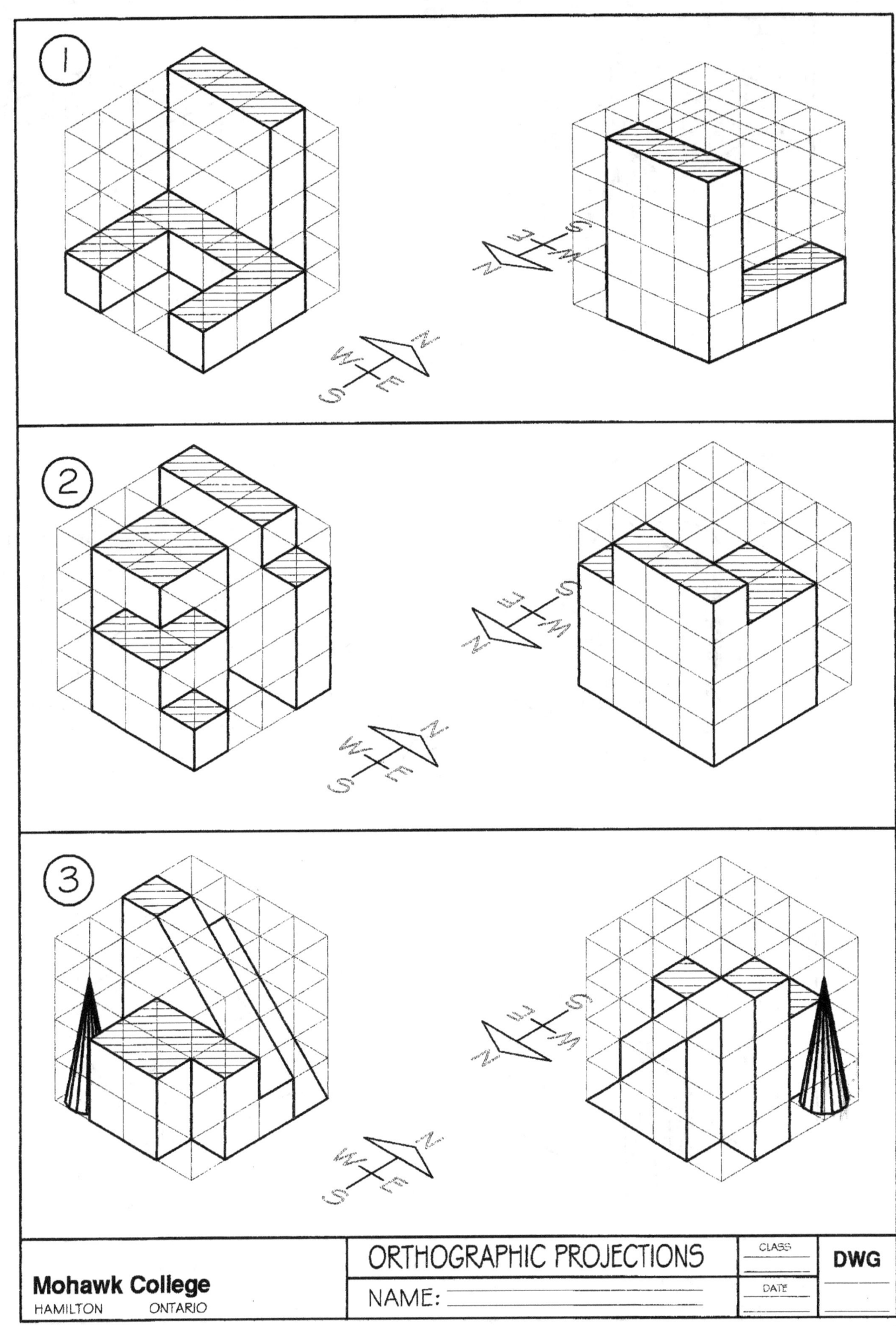

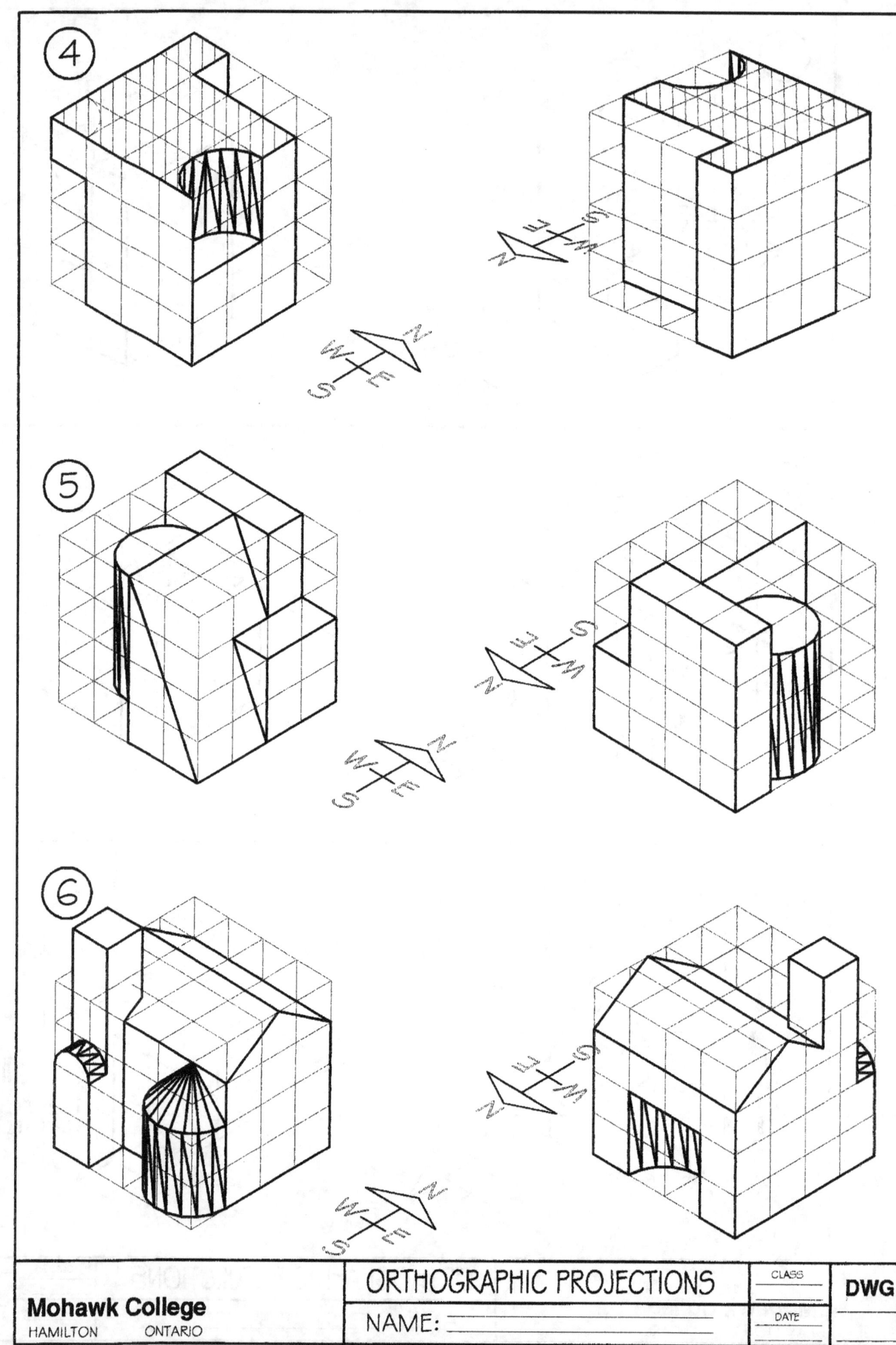

①

| TOP PLAN VIEW | BOTTOM PLAN VIEW | SOUTH ELEVATION |

| EAST ELEVATION | NORTH ELEVATION | WEST ELEVATION |

②

| TOP PLAN VIEW | BOTTOM PLAN VIEW | SOUTH ELEVATION |

| EAST ELEVATION | NORTH ELEVATION | WEST ELEVATION |

Mohawk College
HAMILTON   ONTARIO

ORTHOGRAPHIC PROJECTIONS

NAME:

CLASS

DATE

DWG

③

TOP PLAN VIEW   BOTTOM PLAN VIEW   SOUTH ELEVATION

EAST ELEVATION   NORTH ELEVATION   WEST ELEVATION

④

TOP PLAN VIEW   BOTTOM PLAN VIEW   SOUTH ELEVATION

EAST ELEVATION   NORTH ELEVATION   WEST ELEVATION

Mohawk College
HAMILTON   ONTARIO

ORTHOGRAPHIC PROJECTIONS

NAME:

CLASS
DATE

DWG

⑤

TOP PLAN VIEW	BOTTOM PLAN VIEW	SOUTH ELEVATION
EAST ELEVATION	NORTH ELEVATION	WEST ELEVATION

⑥

TOP PLAN VIEW	BOTTOM PLAN VIEW	SOUTH ELEVATION
EAST ELEVATION	NORTH ELEVATION	WEST ELEVATION

Mohawk College
HAMILTON   ONTARIO

ORTHOGRAPHIC PROJECTIONS

NAME:

CLASS

DATE

DWG

| TOP PLAN VIEW | BOTTOM PLAN VIEW | SOUTH ELEVATION |

| EAST ELEVATION | NORTH ELEVATION | WEST ELEVATION |

| TOP PLAN VIEW | BOTTOM PLAN VIEW | SOUTH ELEVATION |

| EAST ELEVATION | NORTH ELEVATION | WEST ELEVATION |

Mohawk College
HAMILTON   ONTARIO

ORTHOGRAPHIC PROJECTIONS

NAME:

CLASS

DATE

DWG

| TOP PLAN VIEW | BOTTOM PLAN VIEW | SOUTH ELEVATION |

| EAST ELEVATION | NORTH ELEVATION | WEST ELEVATION |

| TOP PLAN VIEW | BOTTOM PLAN VIEW | SOUTH ELEVATION |

| EAST ELEVATION | NORTH ELEVATION | WEST ELEVATION |

ORTHOGRAPHIC PROJECTIONS

NAME:

CLASS

DATE

DWG

Mohawk College
HAMILTON   ONTARIO

ORTHOGRAPHIC PROJECTIONS - Continued

Assignment 2:

The following are two compositions in an isometric view. They both are "carved out" of a cube. Light lines and grid dots represent the 3D cube grid of 5X5 units.

Visualize two 3D sections A-A and B-B for each composition. Remove the work sheet from your manual. Draw the two sections per each composition on the attached assignment sheet. Your Instructor will request freehand or drafted sheets to be handed in.

Assignment 3:

Build three-dimensional models, out of Styrofoam, for compositions in Assignment 1 and 2. Models to be completed will be selected by the Instructor.

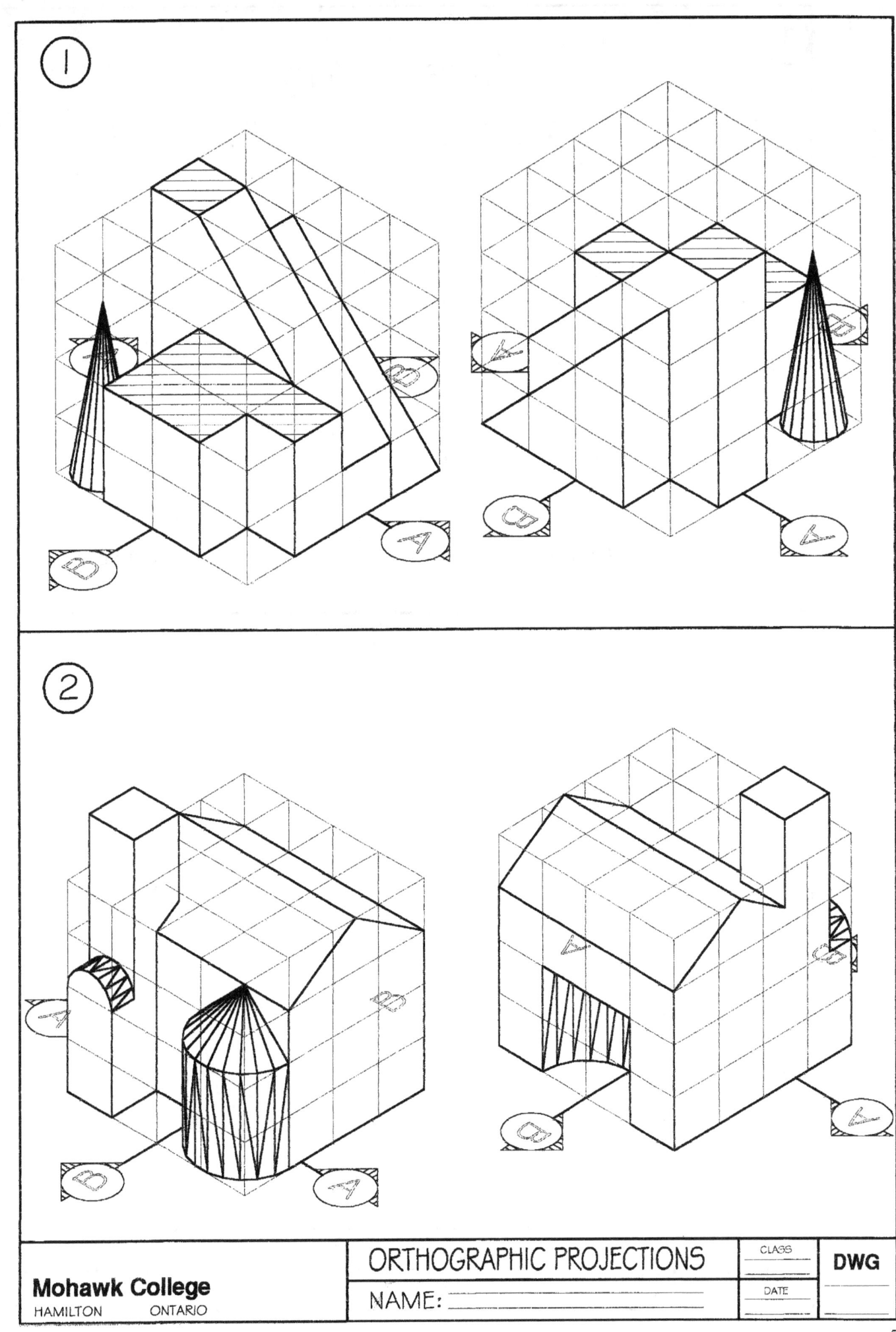

ORTHOGRAPHIC PROJECTIONS

Mohawk College
HAMILTON ONTARIO

ASSIGNMENT

①

SECTION A-A   SECTION B-B

②

SECTION A-A   SECTION B-B

ORTHOGRAPHIC PROJECTIONS	CLASS	DWG
NAME:	DATE	

Mohawk College
HAMILTON   ONTARIO

ASSIGNMENT

SECTION A-A   SECTION B-B

SECTION A-A   SECTION B-B

| Mohawk College | ORTHOGRAPHIC PROJECTIONS | CLASS | DWG |
| HAMILTON ONTARIO | NAME: | DATE | |

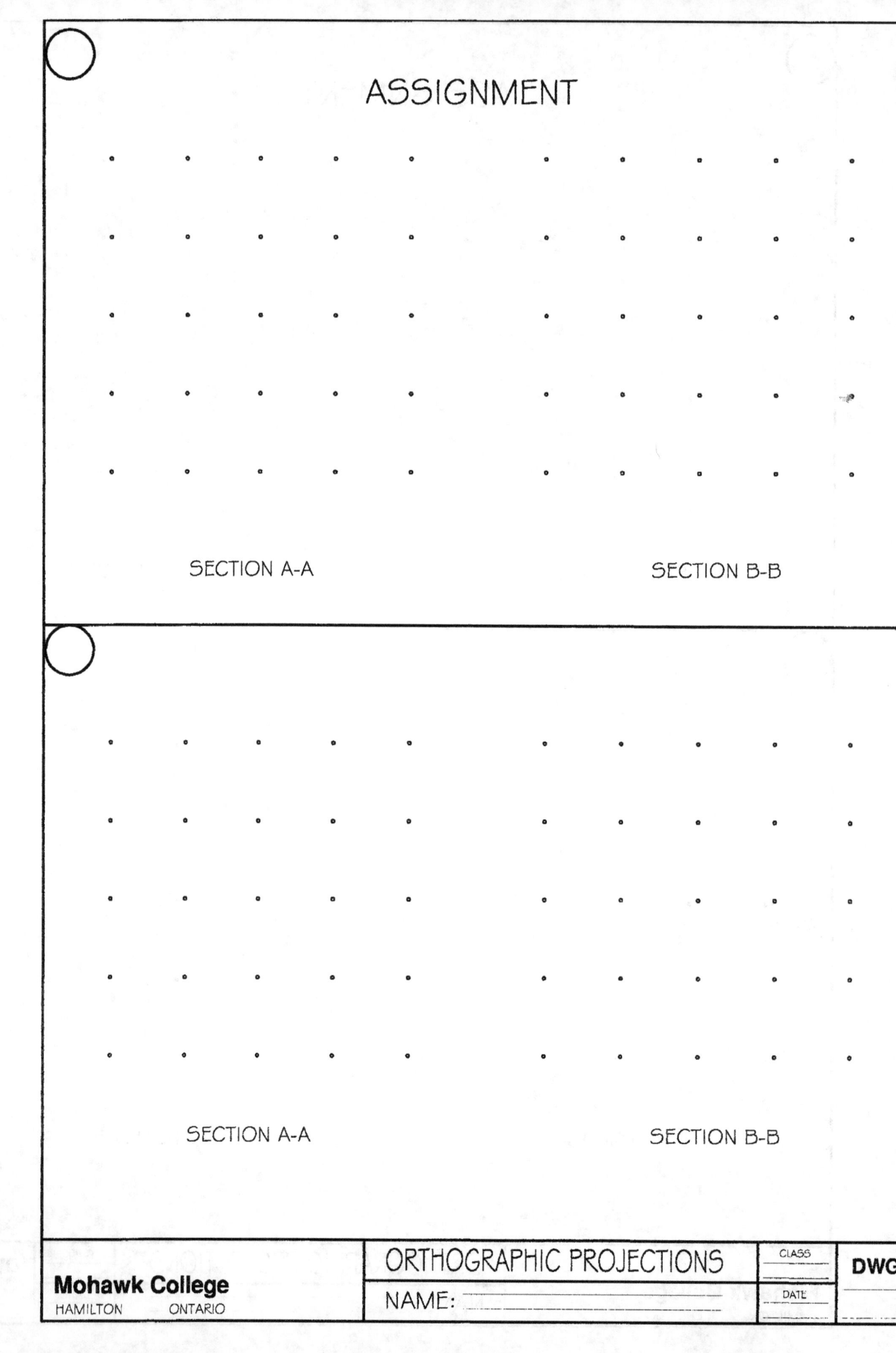

Department of
BUILDING AND CONSTRUCTION SCIENCES
Mohawk College

ARCHITECTURAL SECTION

Your Instructor may choose assignments from the following

OBJECTIVE MODULES:

ADVANCED FREEHAND DRAWING

ADVANCED GEOMETRY

LANDSCAPING PROJECTS

FURNITURE AND FIXTURE LAYOUTS

HOUSES:
Plans, Elevations, Sections, Details

CONSTRUCTION DETAILS

When a job is handed to me I tuck it away in my memory, not allowing myself to make any sketches for months on end. That's the way the human head is made: it has certain independence. It's a box into which you can toss the elements of a problem any which way and leave it to "float", to "simmer", to "ferment". Then one day there comes a spontaneous movement from within, the catch is sprung; you take a pencil, a drawing charcoal, some color pencil's (color is the key to the maneuver) and you give it birth on a sheet of paper. The idea comes out – the child comes out, it comes into the world, it is born.

Le Corbusier

It is suggested that architectural students study the work of the following architects:
Modern history:
- Frank Lloyd Wright
- Le Corbusier
- Mies van der Rohe
- Alvar Aalto
- Philip Johnson
- Lois Khan

Contemporary American Architects:
- Arquitectonica
- Peter Eisenman
- Frank O. Gehry
- Helmut Jahn
- Richard Meier
- Ieoh Ming Pei
- Robert Ventury
- Cesar Peli

Contemporary European Architects
- Dante Benini
- Arto Sipinen
- Ricardo Bofill
- Troughton McAslan
- Adrien Fainsilber
- Antoine Grumbach

Contemporary Canadian Architects:
- Kuwabara Payne McKenna Blumberg Architects
- Patakau Architects
- Saucier + Perrotte Architects

ADVANCED FREEHAND DRAWING INSTRUCTIONS
This segment is optional depending on your Instructor.

An artistic portfolio is required for enrollment in any post-secondary Architecture Program. Most architectural firms require some degree of artistic abilities from their employees.

Proficiency and excellence in freehand drawing can be achieved through practice. Anything can be drawn. Build a composition about 15-20ft in front of you Draw light, continuous lines first. Establish the overall proportions of your composition. Rotate your board accordingly. Compose your drawing so there is a fair margin space all around but draw your composition large enough to fill out the sheet. The examples shown on the next page were drawn on 11"X17" sheets mounted on hardboards.

Start with simple geometrical boxes. Measure and compare each box proportion as well as the angle of lines to be shown in perspective. Establish a horizon line at exactly your eye level. Locate imaginary vanishing points for all parallel lines. Put your board aside often in order to examine your progress. Compare your proportions and keep correcting your drawing. Continue to draw long, continuous lines; overlay many lines on each other. This will give your drawing the freehand feel and expression. The final result should emerge darker as you apply more lines. Do not draw very heavy, final lines. These will only destroy and trivialize your efforts.

"Box out" all complex components of your composition; enclose them within imaginary boxes. Draw these boxes first by establishing the right proportions. Enclose with such an imaginary box, elements like plants, trees, and parts of a human body. Fill in more and more straight, measured lines and gradually develop the complex shapes. Add more difficult elements to your compositions. Complete your training with a live drawing of a person. This is most difficult. Any mistake in proportion and your model will not look human. You should be able to reflect the person's size, age, type of hair and the clothing worn. When successful with this final exercise, you will be able to draw any structure to its true proportions, and capture the structures' materials and details.

There are many measuring techniques and drawing practices. Your instructor will offer demonstrations or textbook references.

Dorota Goede B.Arch

Mohawk College
HAMILTON   ONTARIO

FREEHAND SKETCHING
ADVANCED INSTRUCTIONS

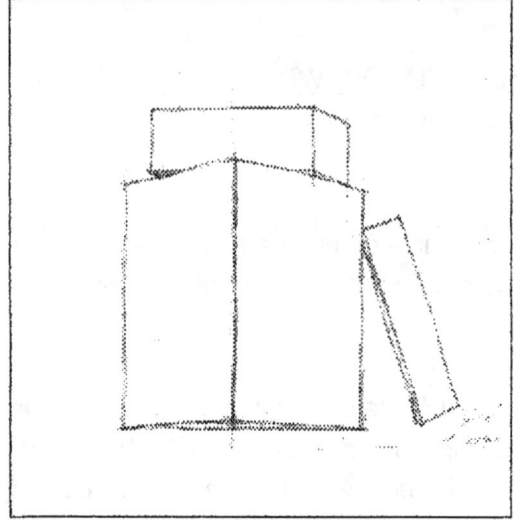

GEOMETRIC BOXES AND PROPORTIONS
PARALLEL LINES AND PERSPECTIVE

CYLINDRICAL SHAPES, ELLIPSES
SURFACE SHADING

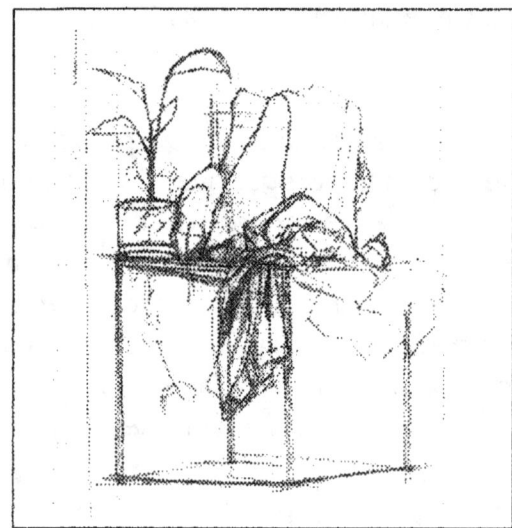

FABRICS, PLANTS AND SHOES.
MATERIALS, TEXTURES, SURFACES & FOLDS

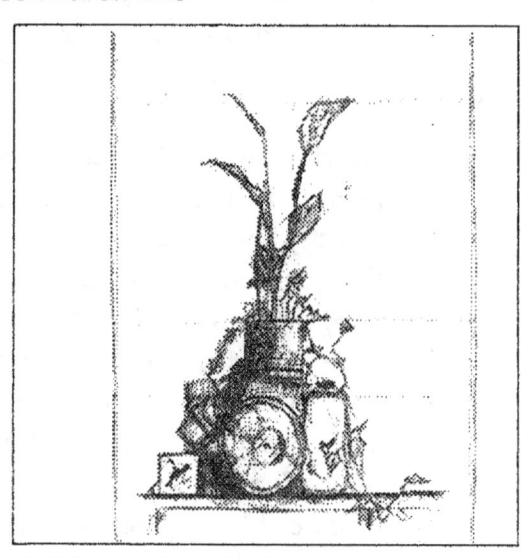

COMPOSITION OF GEOMETRIC SHAPES.
EXERCISE IN INK PEN

FIGURE DRAWING; EXERCISE IN PROPORTION, GEOMETRY, MATERIALS & COMPOSITION.

Dorota Goede B.Arch

| Mohawk College HAMILTON ONTARIO | FREEHAND PRACTICE STAGES | CLASS 2AR23 | DWG |
| | Architectural Technology Students | DATE 1992 | |

CONDOMINIUM FURNITURE LAYOUT PROJECT

Your Client has just replaced old carpets with a new ceramic tile in her condominium. She asks you to design a furniture layout to assist her in the next renovation stage. The Condominium Floor Plan, compiled for the tile layout, is made available to you. The plan, as shown on the next page, is printed in exactly 1/4" = 1'-0" scale.

ASSIGNMENT 1: Trace freehand the furniture shown to 1/4" scale below. Propose two layouts on attached sheets. Render (colour) your proposals for presentation to a Client.

ASSIGNMENT 2: Trace the plan on A2 size vellum sheet twice. Propose two alternative layouts using typical furniture templates below. Use the templates by tracing furniture onto your plans. You may want to do some freehand sketches and proposals first. Not all furniture must be used.
When completed, retape the handout in the original location and trace the tile lines very lightly. Do not show tile over your furniture! Add all doors and indicate door swings.
The final stage of your project involves preparation of your proposal for your Client review meeting. Use pencil crayons and highlight the solid walls, furniture, bedroom carpet, tile etc. If you do not feel confident about your technique, try samples on a blueprint, or make your proposal on the backside of velum. Use colour pencils lightly; apply each colour in light layers. Use only two or three colours; pastels are always a "safe" choice. Have fun!

Colour-pencil rendering technique is very popular in architectural offices. It allows for very quick and effective presentation. Different colours can represent a number of aspects of the proposal. For example, corridors and exits are usually rendered red for Ontario Building Code review meetings with the town officials.

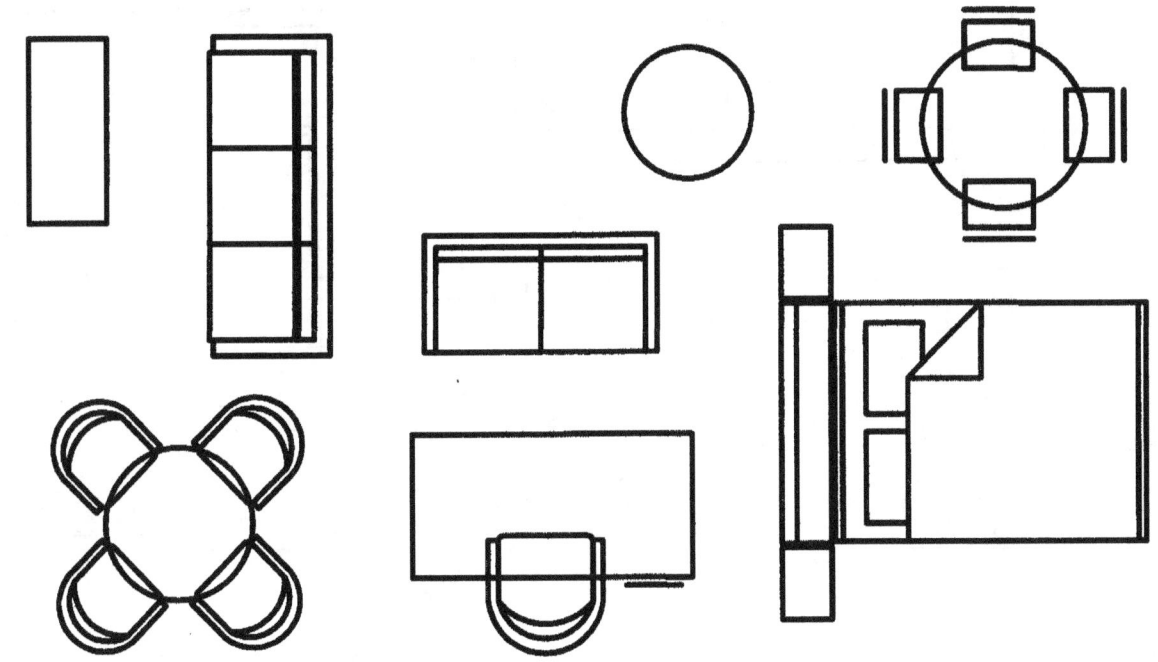

Mohawk College
HAMILTON   ONTARIO

CONDOMINIUM LAYOUT
INSTRUCTIONS

CLASS
DATE
DWG

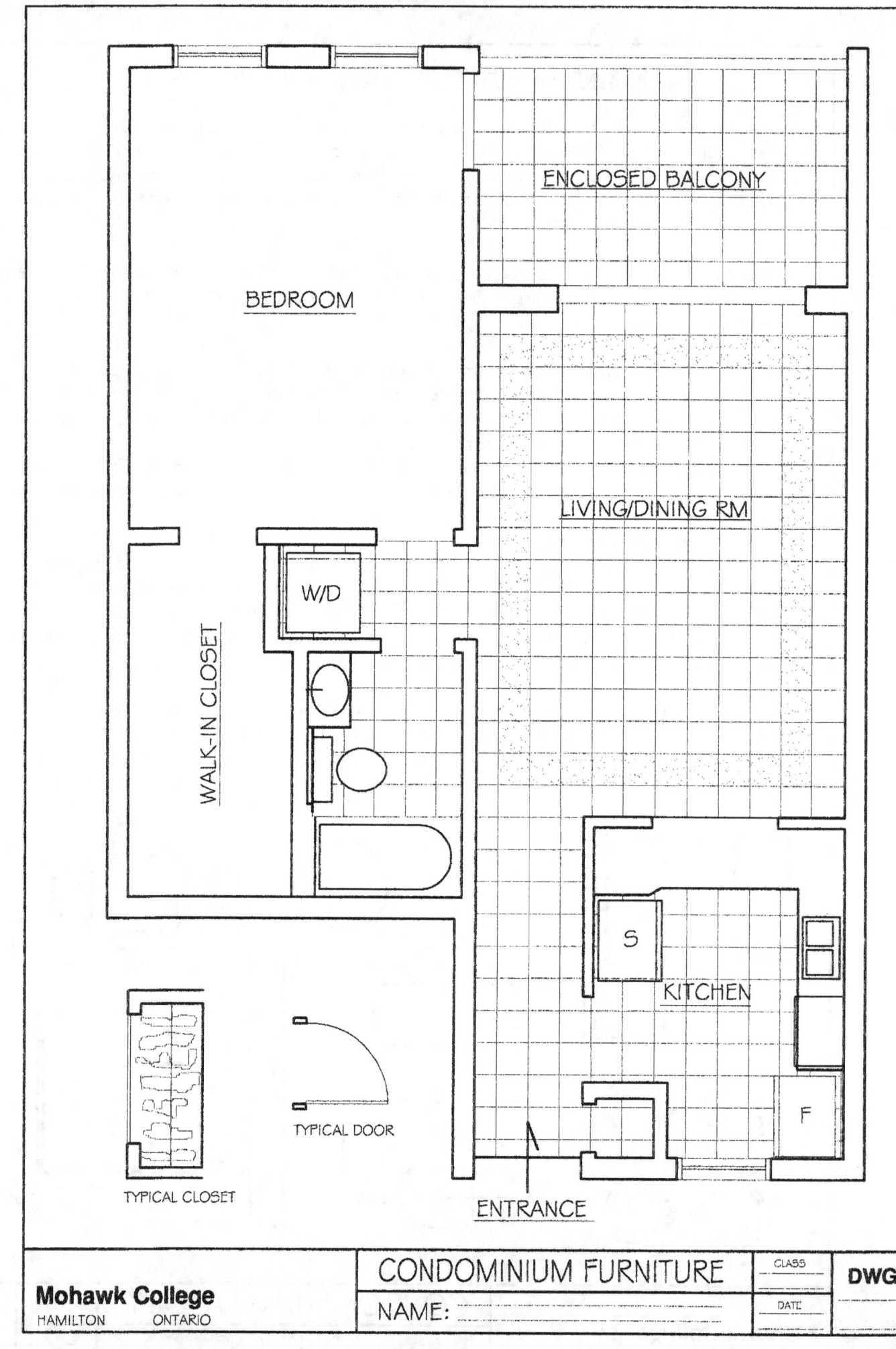

ROOF GEOMETRY CONSTRUCTION ASSIGNMENT

Assignment 1: Draw the roof outline with dark lines using the dimensions shown. Construct the hip roof structure and geometry using the 45/45/90 triangle. Offset the dashed wall outline by a 12" soffit. Construct locations for six, 12" diameter garden columns on the North side. Use a geometric line division technique to establish the precise and equal line division into 5 equal spaces between the columns.
Hatch parallel roof surfaces as shown. Use very light horizontal and vertical lines and draw them using consistent spacing between the lines.
Use A2 size sheet and draw your project in 1/4" = 1' - 0", architectural scale.

Assignment 2: Design and draw the North garden patio and the front entrance driveway and walkways. Design landscaping. (Optional for Architectural students).

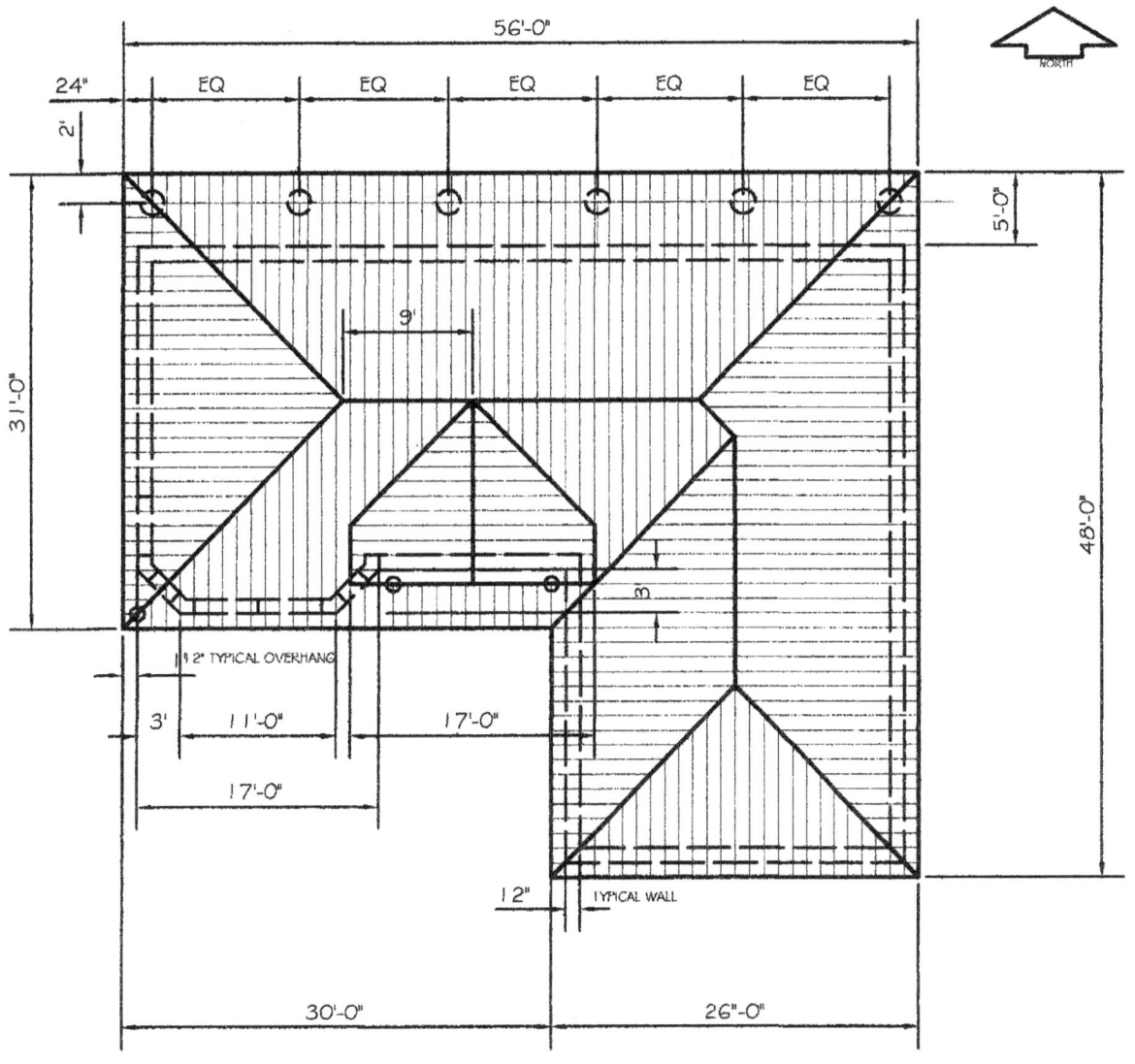

Dorota Goede B.Arch

CREANONA ROOF PLAN
ASSIGNMENT

Mohawk College
HAMILTON   ONTARIO

GARDEN SHED FRAME DETAILS PROJECT INSTRUCTIONS.

Draw the simple garden shed framing plan and elevations as shown on following pages. Use A2 size sheet. Drawing A1 will contain Floor Plan in 1/2"=1'-0" scale. A2 will contain shed elevations. You are asked to research and draw your own wall sections showing materials you propose for your own shed.

The objective of this exercise is to produce typical construction drawings as attached with installation guide sheet when purchasing a construction kit for a simple garden structure. Your Instructor may change or add to the project.

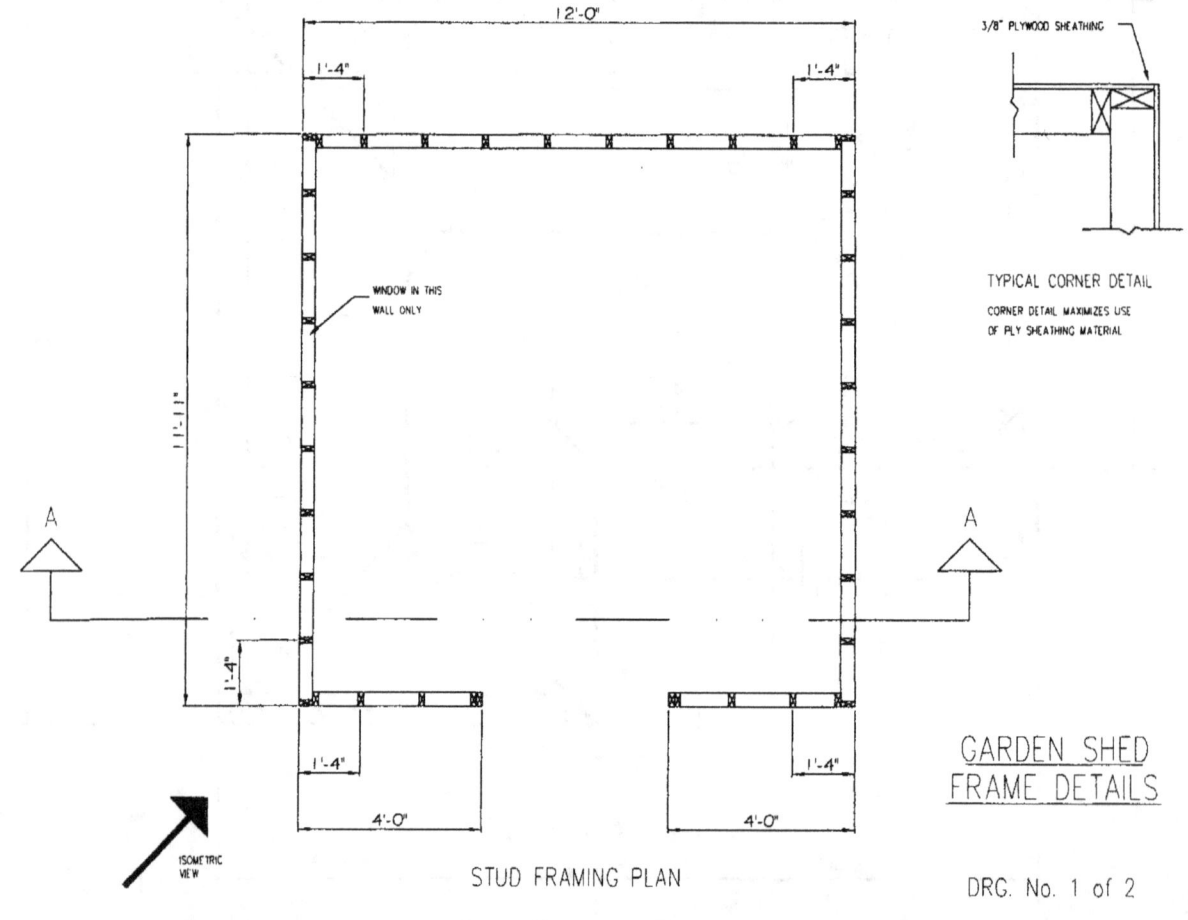

STUD FRAMING PLAN

TYPICAL CORNER DETAIL
CORNER DETAIL MAXIMIZES USE OF PLY SHEATHING MATERIAL

GARDEN SHED FRAME DETAILS

DRG. No. 1 of 2

MANN DEC. 1992

Mohawk College
HAMILTON   ONTARIO

GARDEN SHED
FRAMING PLAN

SCALE: 1/4"=1'-0"
DATE: Nov.02'

DWG **A1**

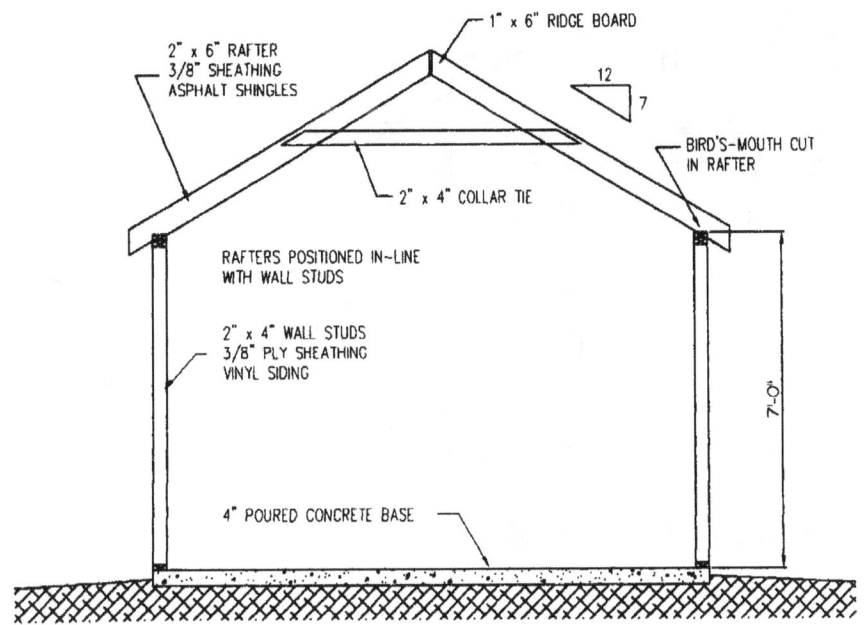

CROSS SECTION A-A

ADDITIONAL ASSIGNMENT INSTRUCTIONS:

USING THE FRAMING DETAILS, DRAW AN ISOMETRIC VIEW OF THE SHED FROM THE DIRECTION OF THE ARROW SHOWN ON THE PLAN VIEW. SHOW ALL STRUCTURE MEMBERS INCLUDING THE CONCRETE BASE

GARDEN SHED FRAME DETAILS

DRG. No. 2 of 2

MANN DEC. 1992

Mohawk College
HAMILTON ONTARIO

GARDEN SHED

NAME:

CLASS 1/4"=1'-0"
DATE Nov.02'

DWG **A2**

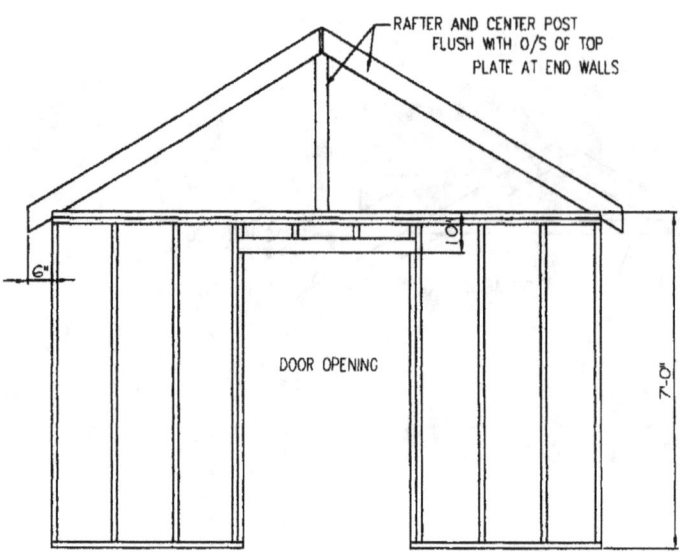

FRONT ELEVATION WITH DOOR

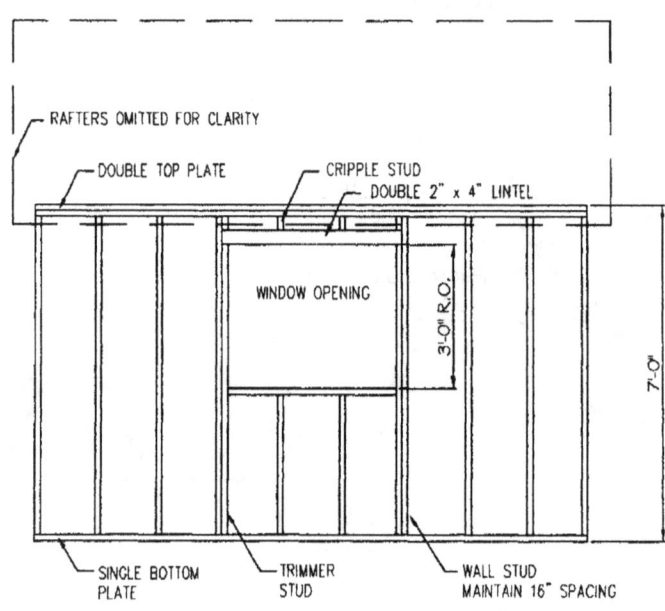

TYPICAL WALL ELEVATION
SEE PLAN FOR WINDOW LOCATION

HOUSE GARAGE PROJECT

Design and draw floor plan, section and wall section for a typical house garage. Use the following pages as an example of type of information required for construction. Draft on A2 size sheet and decide on most suitable imperial scale for this project.

Start your project from designing your sheet layout and locate your titleblock. Outline all drawings lightly with construction lines to ensure enough room for all components. The suggested scale is ½" = 1'0"
Draw all exterior walls and roofs darkest. All elements shown in section must also be drafted with the darkest line. Use medium line weight for all dimensions and notes. Ensure that all your construction and guidelines are very precise and light.

The attached two typical wall sections are examples to be used as a drafting standard guide only. You are asked to research and draw your own wall sections showing materials you propose for your garage.

The objective of this exercise is to draw a comprehensive set of construction notes complete with dimensions, notes and material specifications for a small garage.

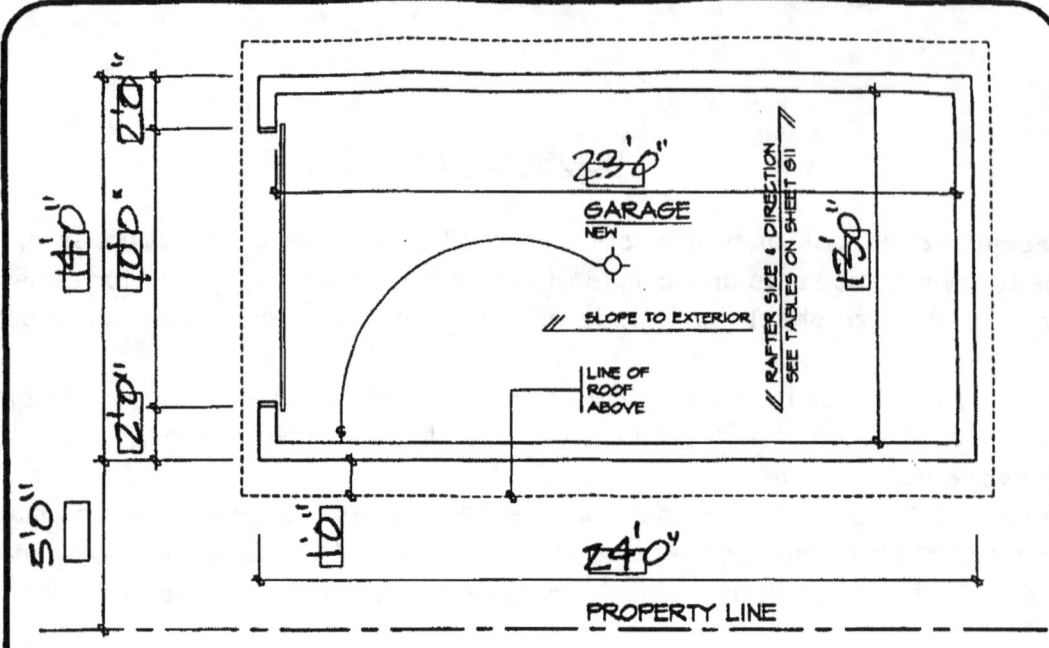

GARAGE PLAN

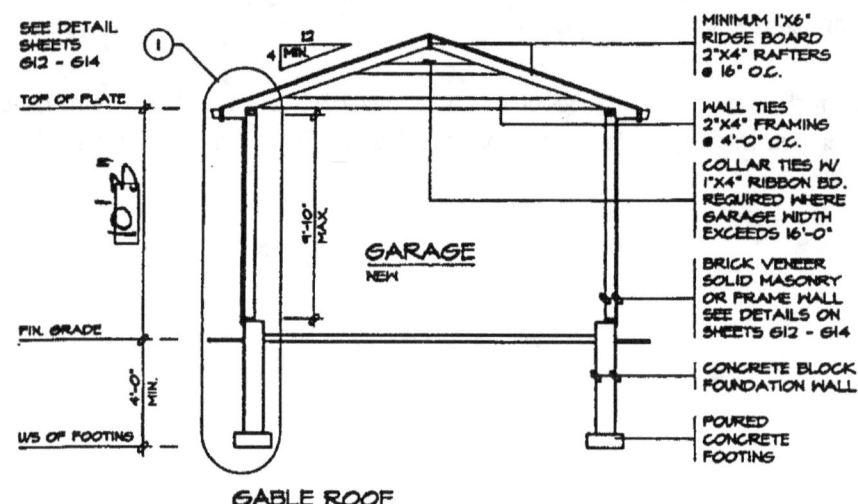

GABLE ROOF

GENERAL NOTES

1. ALL LUMBER TO BE NO. 1 & 2 SPRUCE OR BETTER
2. ALL PLYWOOD SHALL BE STAMPED EXTERIOR GRADE
3. ROOF LOAD DESIGN 21 LB./SQ. FT. OR 31 LB./SQ. FT.
4. ALL FOOTINGS TO BEAR ON UNDISTURBED SOIL.
5. IF GARAGE WALL IS LESS THAN 2'-0" TO THE PROPERTY LINE PROVIDE 5/8" TYPE 'X' DRYWALL INTERIOR SHEATHING. NO WINDOWS ARE PERMITTED IN GARAGE WALLS LESS THAN 3'-11" FROM PROPERTY LINE.
6. FOR ONE STOREY WOOD FRAME DETACHED GARAGES LESS THAN 538 SQ. FT. AN ALTERNATE FOOTING MAY BE USED, SEE DETAIL SHEET G12
7. GARAGE SLAB SHALL BE 4650 PSI CONCRETE W/ 5% - 8% AIR ENTRAINMENT SLOPED TO DRAIN TO THE OUTSIDE.
8. ROOF SHEATHING SHALL BE MIN. 3/8" PLYWOOD PROVIDE 'H' CLIPS IF RAFTERS OR JOISTS ARE SPACED GREATER THAN 16" O.C.
9. PROVIDE A LIGHT FIXTURE IN THE GARAGE.

DETACHED GARAGE	PERMIT APPLICATION NO.	REVIEWED BY:	DWG. NO.
SLOPING (PLAN & SECTIONS		DATE:	

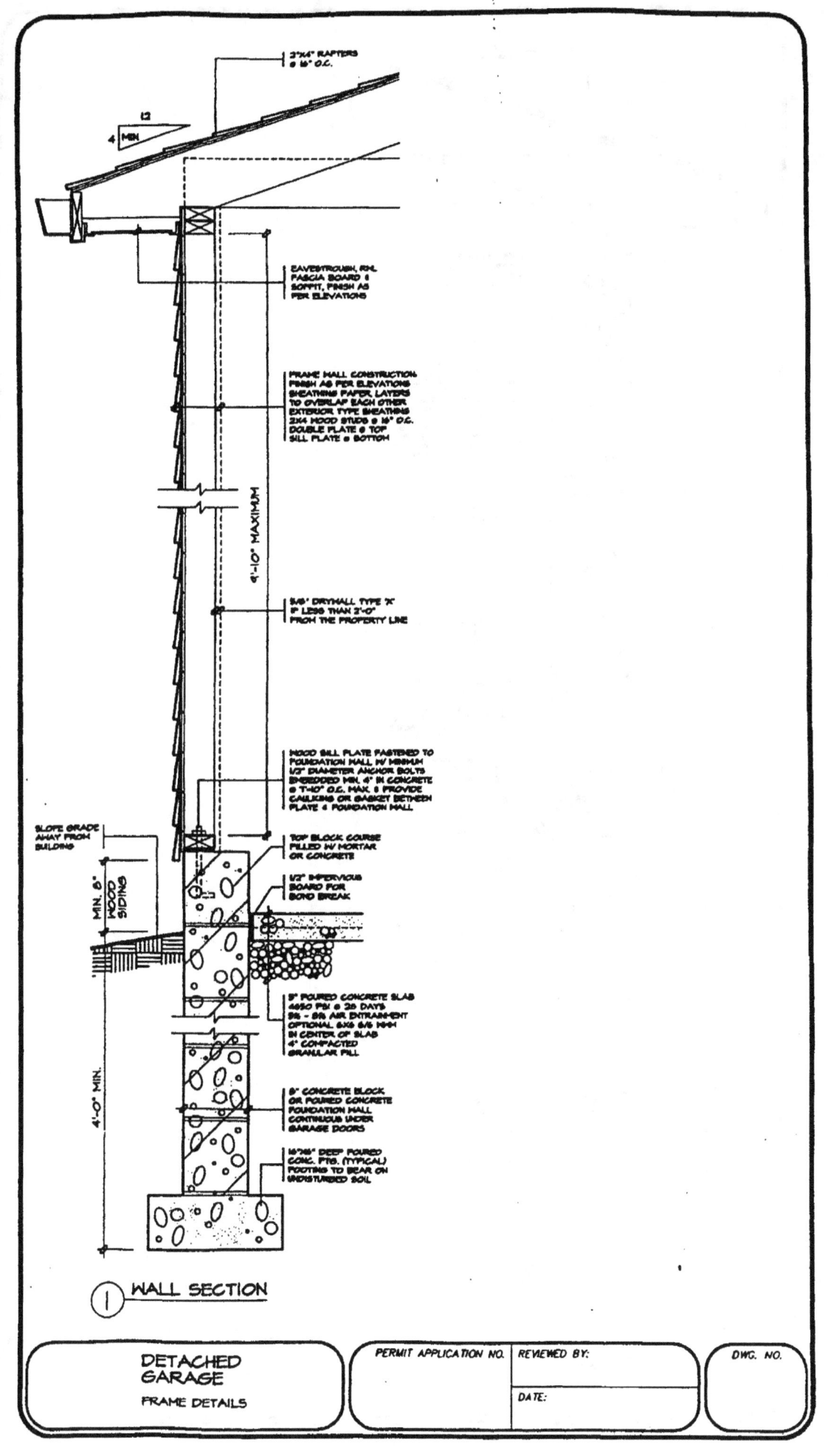

WALL SECTION

DETACHED GARAGE
FRAME DETAILS

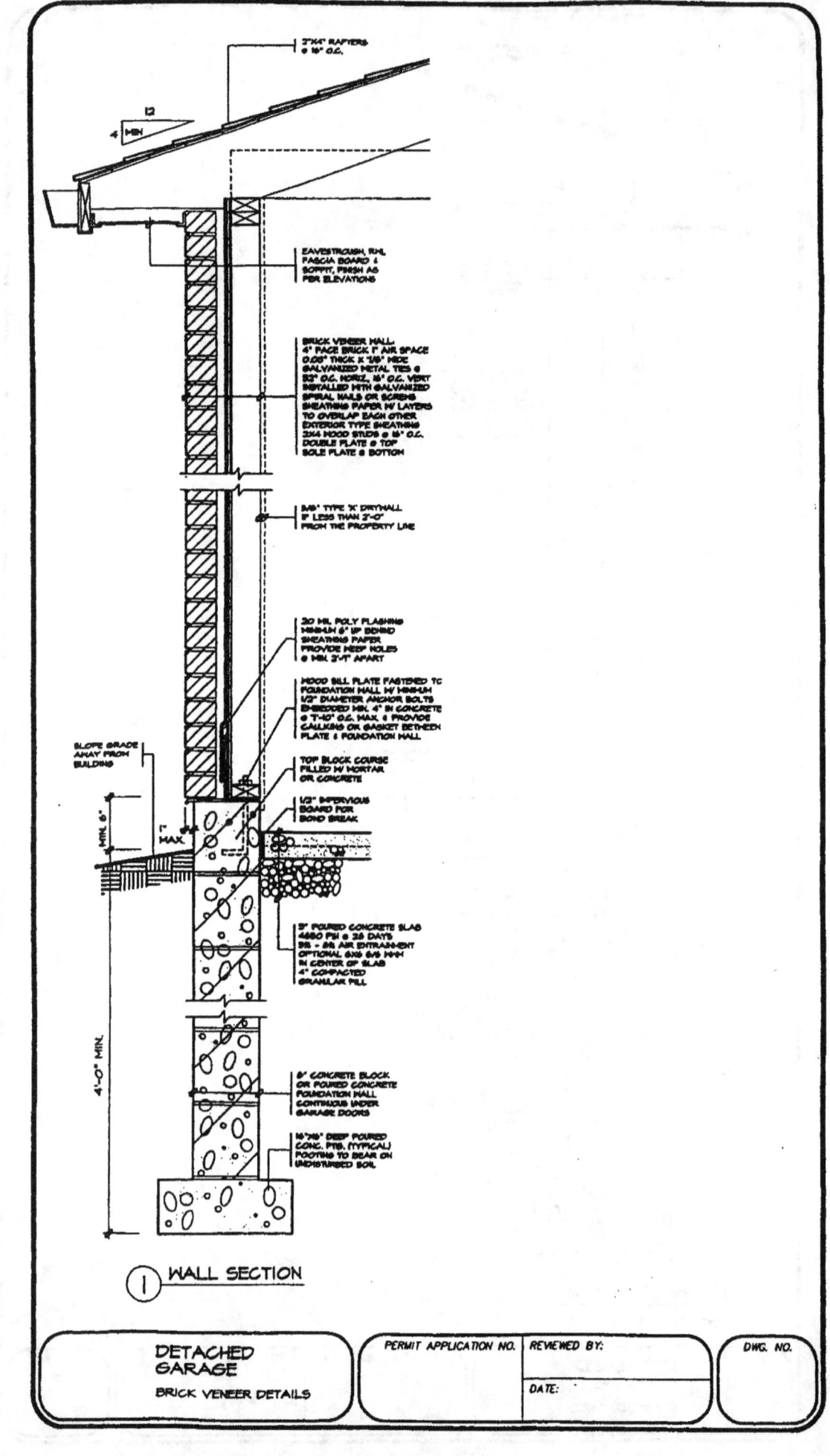

Office Building

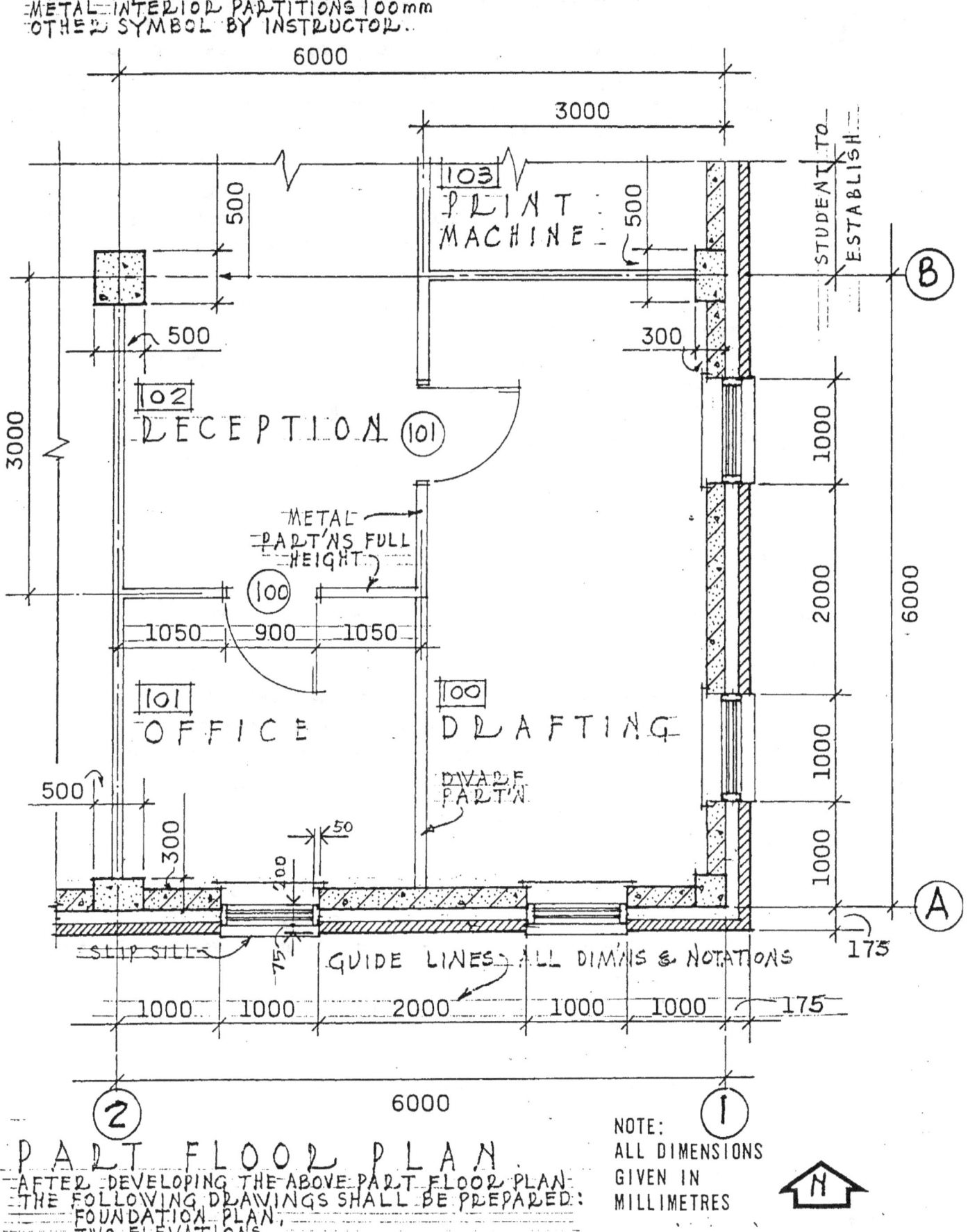

ARCHITECTURAL FINAL PROJECT

PARK ROAD HOUSE DESIGN DEVELOPMENT

The handout sheets, A1-A3, represent the up-to-date residential project development drawings. Your assignment is to continue with the project development stage by proposing the front and garden side elevations. As part of your proposal you must design a roofline for this house. The Site Plan drawing is printed in 1/8"=1'-0" scale and it can be traced. This drawing does not contain all landscaping dimensions so remove the page from your book, tape it on board, and trace on vellum. The roofline, shown with dashed line, can be altered in order to represent your own roofline proposal. Some students may choose to accept a given design.

The hidden (dashed) line on all drawings represents roofs above or fences. The dash-dot line represents property line and structural beams above.

Floor plans are printed not to scale (N.T.S.) and you must draw new plans in 1/4"=1'-0", standard, residential scale.

All interior walls and partitions are 4" thick and all exterior walls are 6" thick. Exterior finish is aluminum siding.

Your final submission will include:

Site Plan drawing	1/8"=1'-0"
Main and Upper Floor plans	1/4"=1'-0" scale
Front and garden side elevations	1/4"=1'-0" scale

Notes:

Decide sheet size and layout. Line up both plans on a sheet. Draw plans as outlined in your handout. You may introduce your own changes.

Draw dark ground line for elevations and show 8'-0" deep basement using a uniformly dashed line. Line up Elevations on your sheet. Composition and layout of your submission is very important and will be evaluated.

Pay special attention to quality and precision of your lines. Start with light construction lines and locate all crucial information. Then, show all object lines darkest. For elements in elevation view, centerlines, dimensions and notes, use a medium line weight. All hatching, construction and guidelines must be the lightest. Use minimum of 3 line weights.

Design your own titleblock per Mohawk College standards.

PROJECT DUE DATE: LAST MEETING THIS SEMESTER.

Mohawk College — HAMILTON ONTARIO

PARK ROAD HOUSE

INSTRUCTIONS

SCALE:
DATE: March 01'

DWG

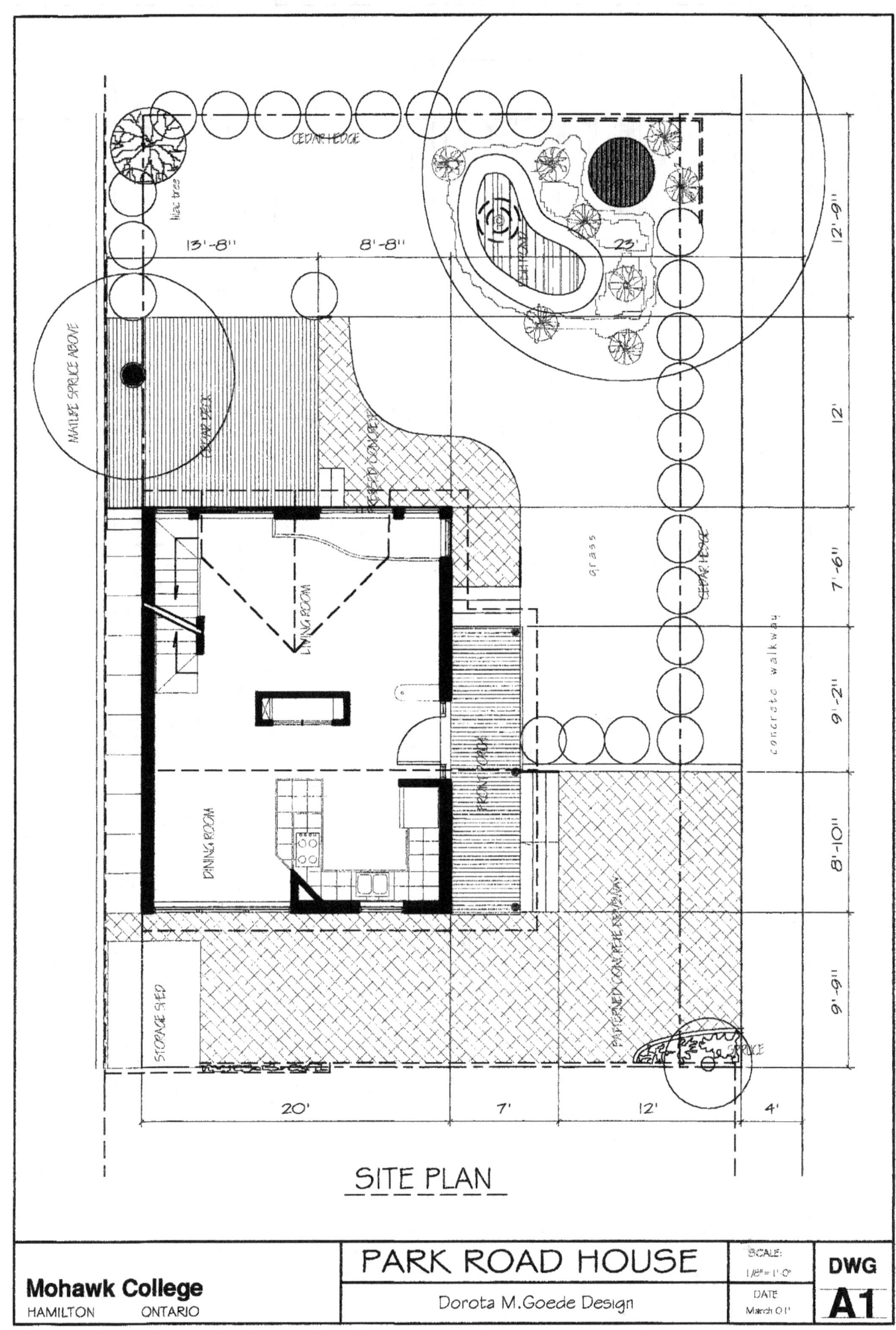

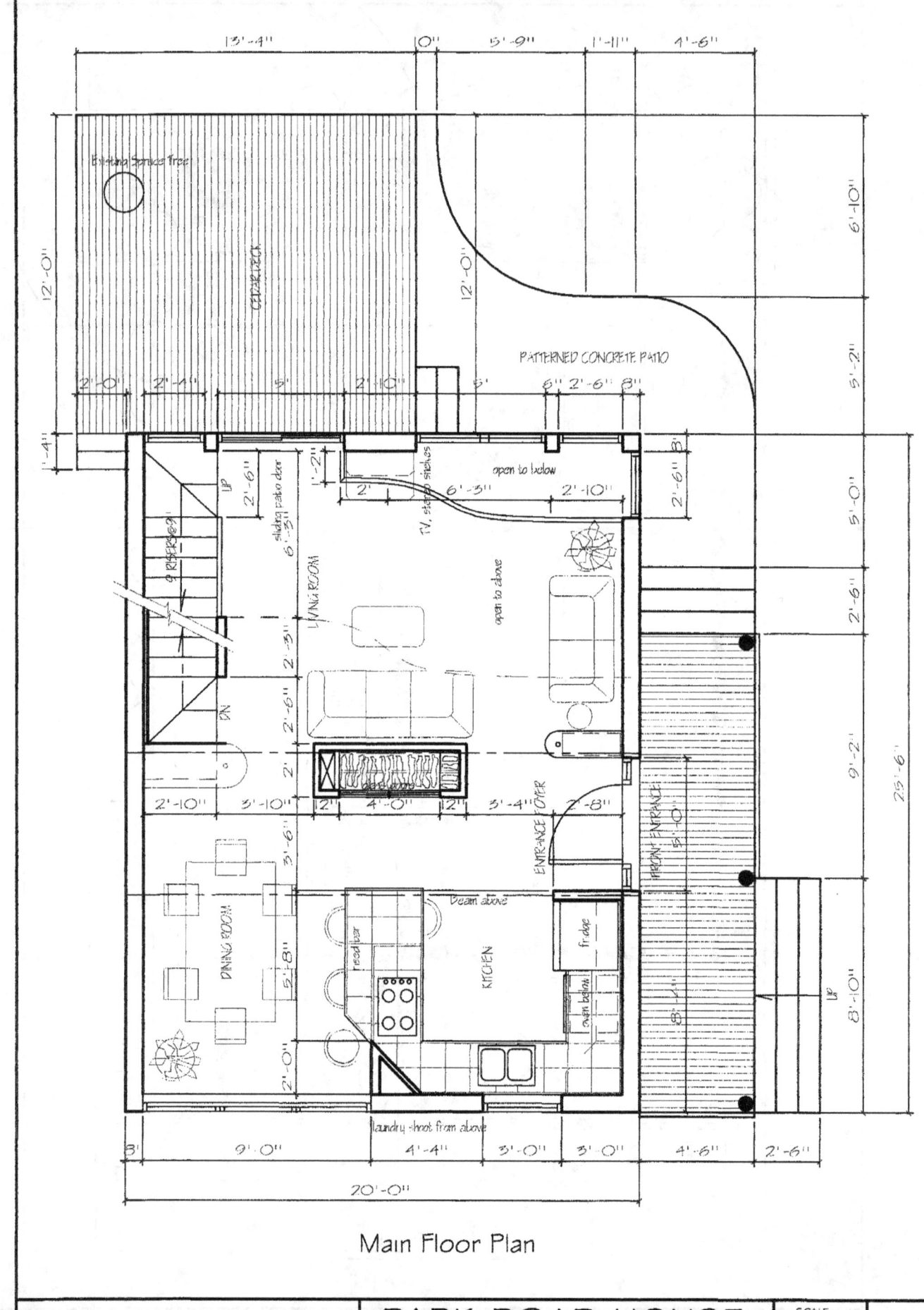

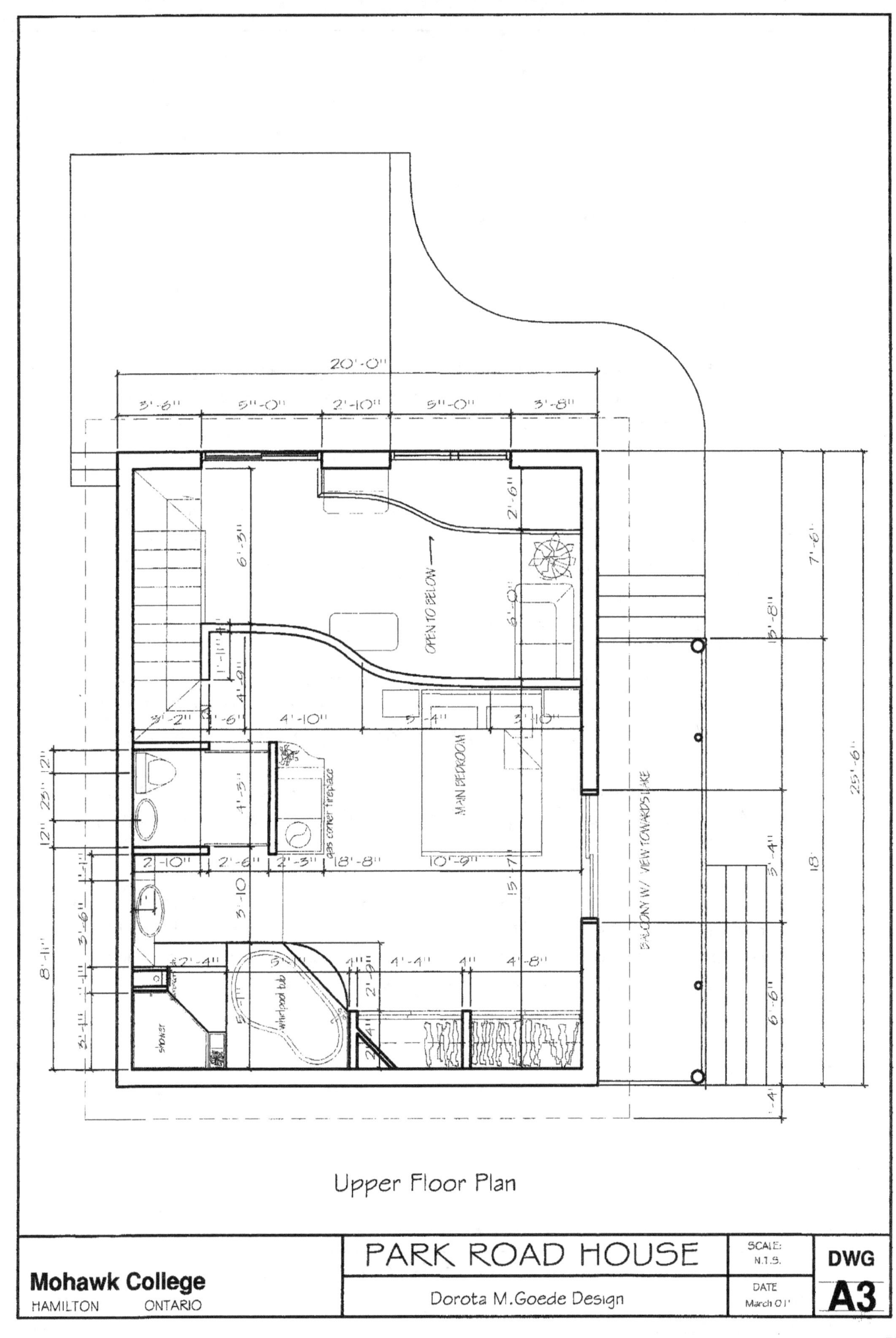

Upper Floor Plan

PARK ROAD HOUSE
Dorota M. Goede Design

Mohawk College
HAMILTON   ONTARIO

SCALE: N.T.S.
DATE: March 01'
DWG A3

CLIENT: Mr. & Mrs. Daniels

BUDGET: $80,000
CLIENT PROGRAM:

- Design and draw oversized double car garage and storage.
- Design new Main Bedroom complete with large bathroom over the garage and connect to the upper level of the existing house.
- Design new front entrance c/w large closet and protected connection to the new garage.
- Improve the front façade. Propose veranda roof tied with the new garage.
- Design new furniture layout and propose work area and guest sleeping accommodations.
- Design future landscape plan c/w new deck, pool and gardens.

This was a real project done quickly using AutoCAD in 1994. It was designed but never constructed per attached plans. A real Client and Program scenario was used, however, for this project.

Trace existing Main and Upper floor plans, and Back and Front Elevations, in ¼"=1'-0" scale. Allow space for a min. 20'X20' garage addition on the South house side.

Our objective is that you research and design your own solutions to this Program and test your abilities in preparation for real life employment. We look forward to seeing your own ideas but some suggestions are included for those of you not knowing "where to start".

Mohawk College
HAMILTON   ONTARIO

FINAL PROJECT

NAME:

CLASS

DATE

DWG

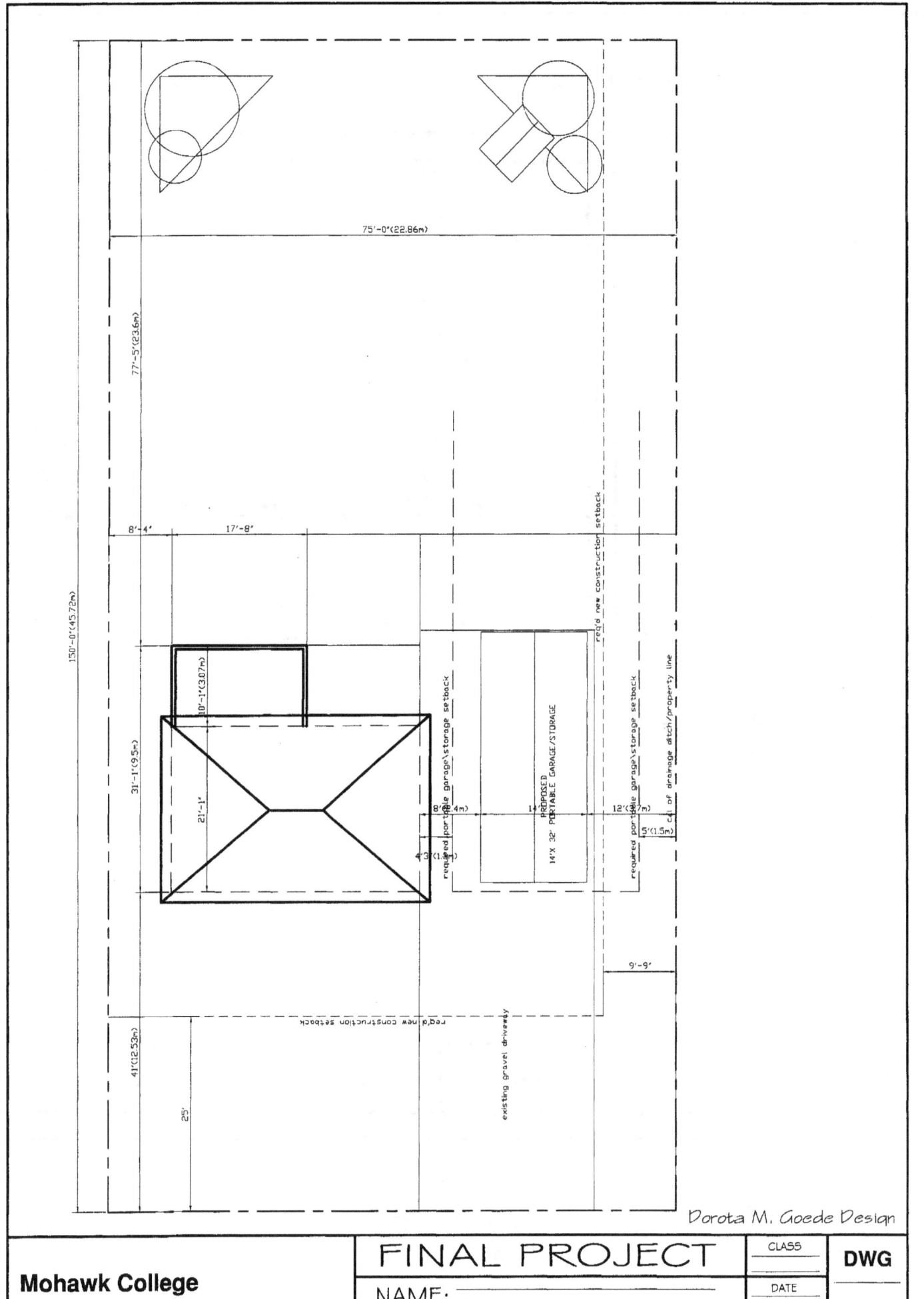

EXISTING MAIN FLOOR PLAN
Scale: 1/4" = 1'-0"

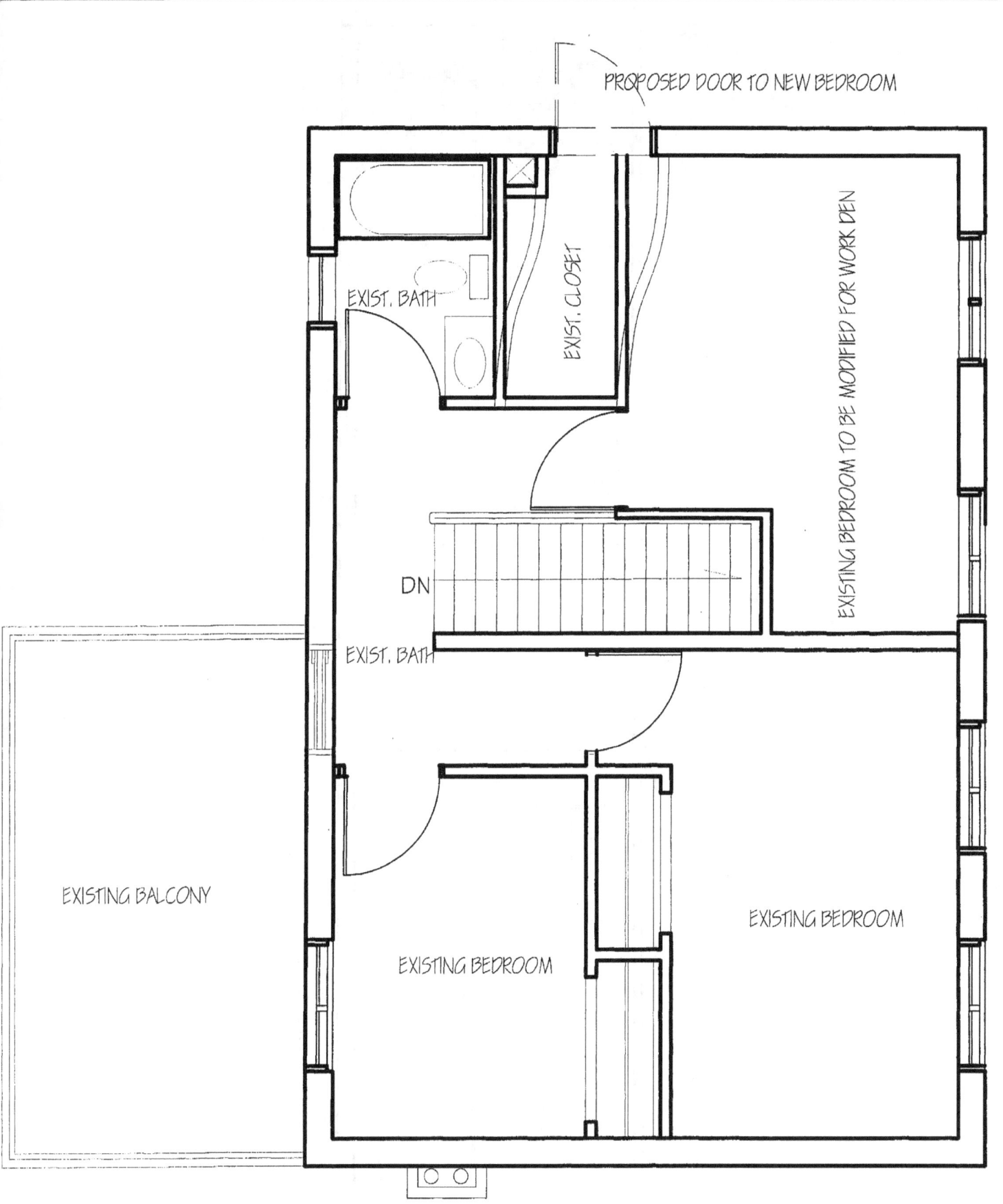

EXISTING UPPER FLOOR PLAN
Scale: 1/4"= 1'-0"

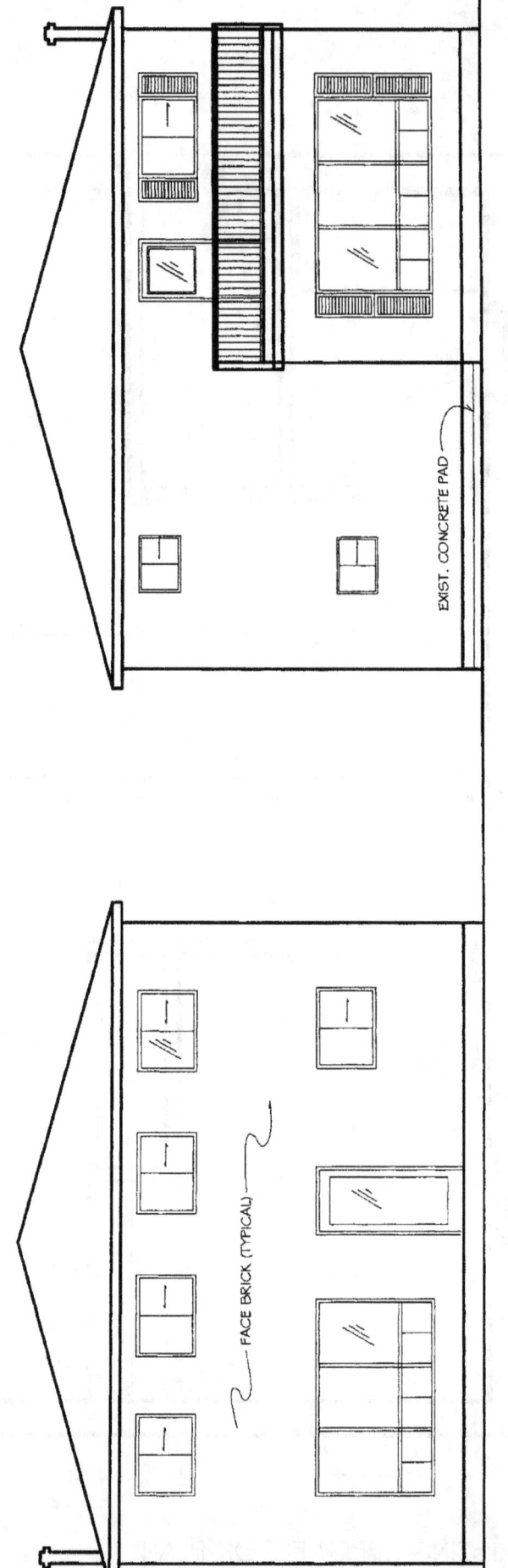

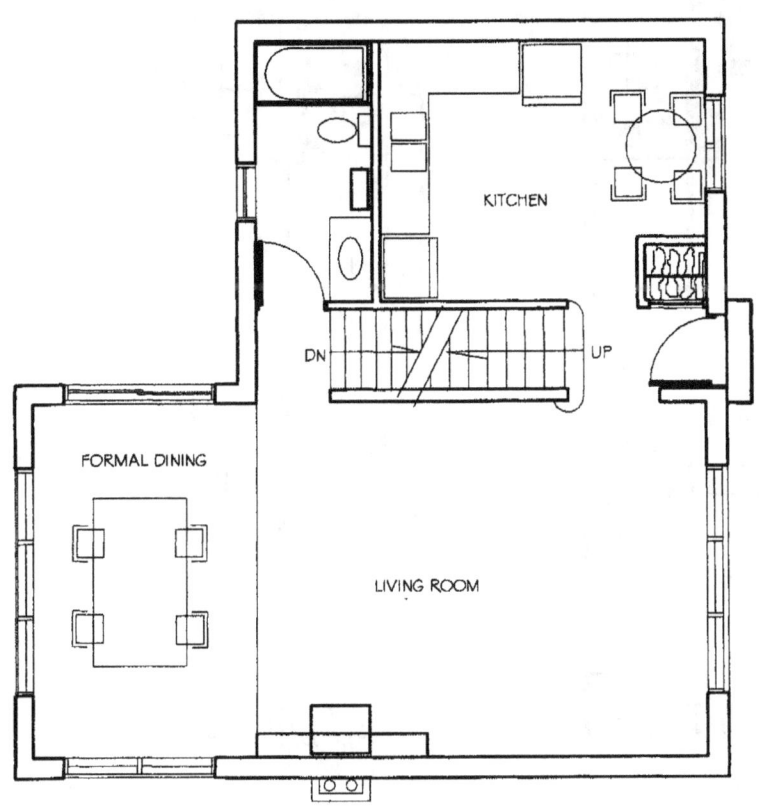

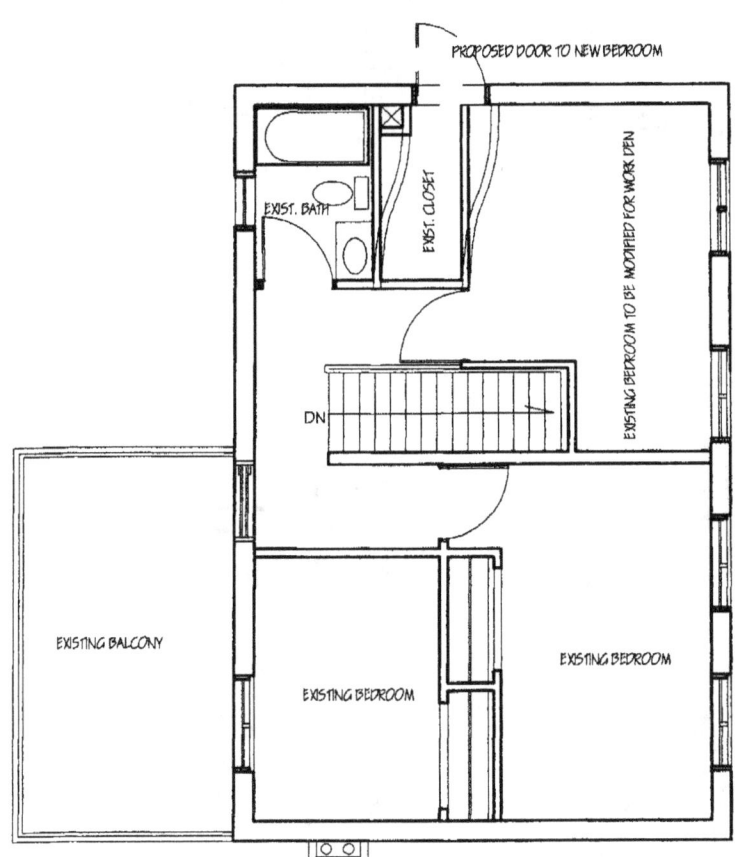

EXISTING FLOOR PLANS
Scale: 1/8" = 1'-0"

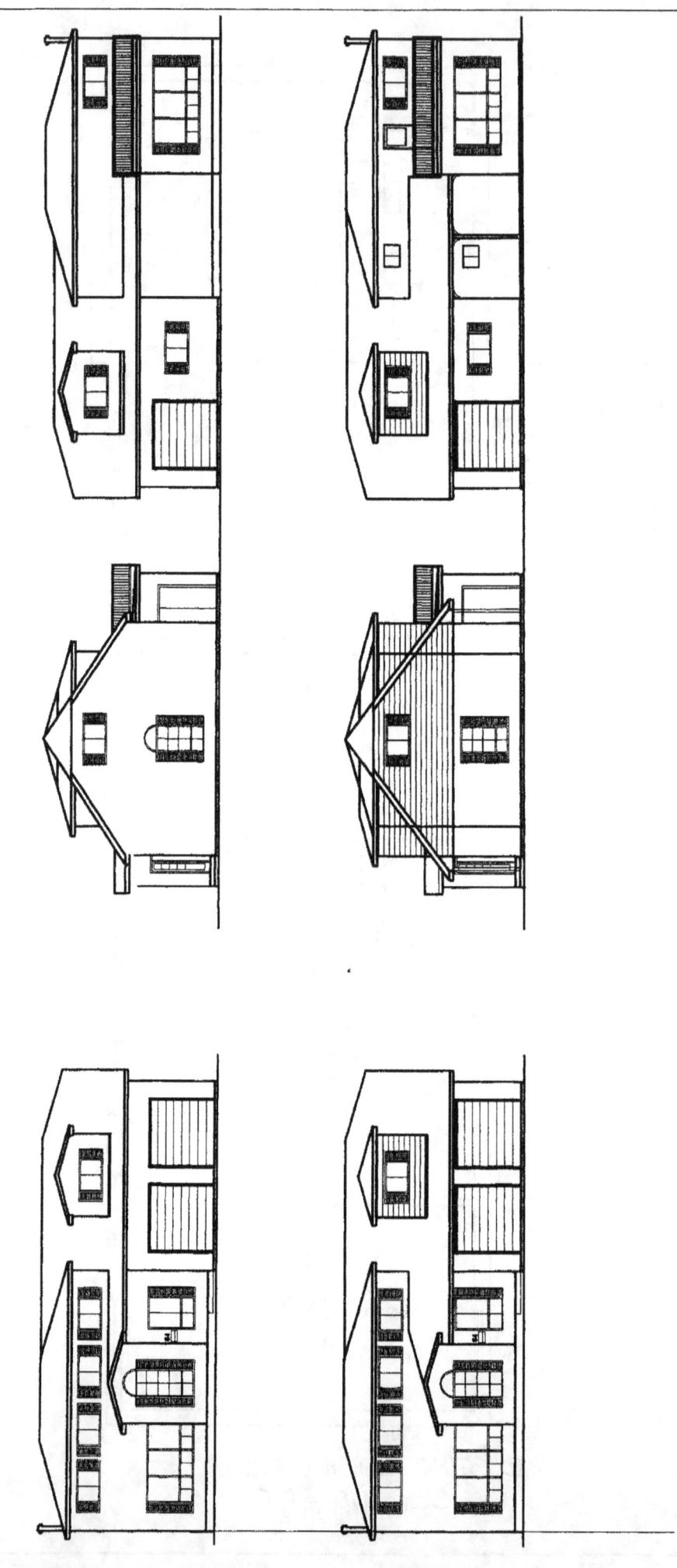

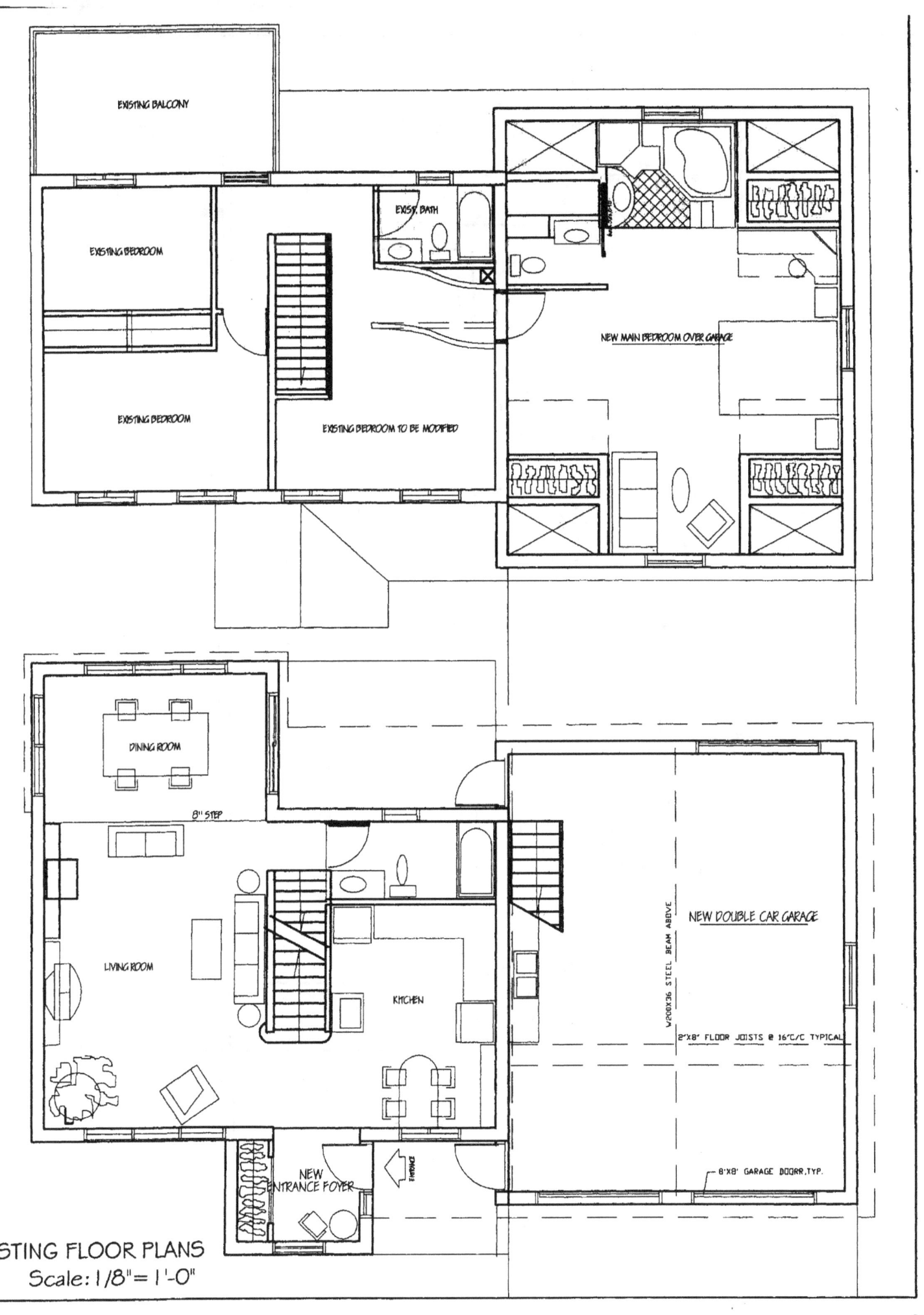

EXISTING FLOOR PLANS
Scale: 1/8" = 1'-0"

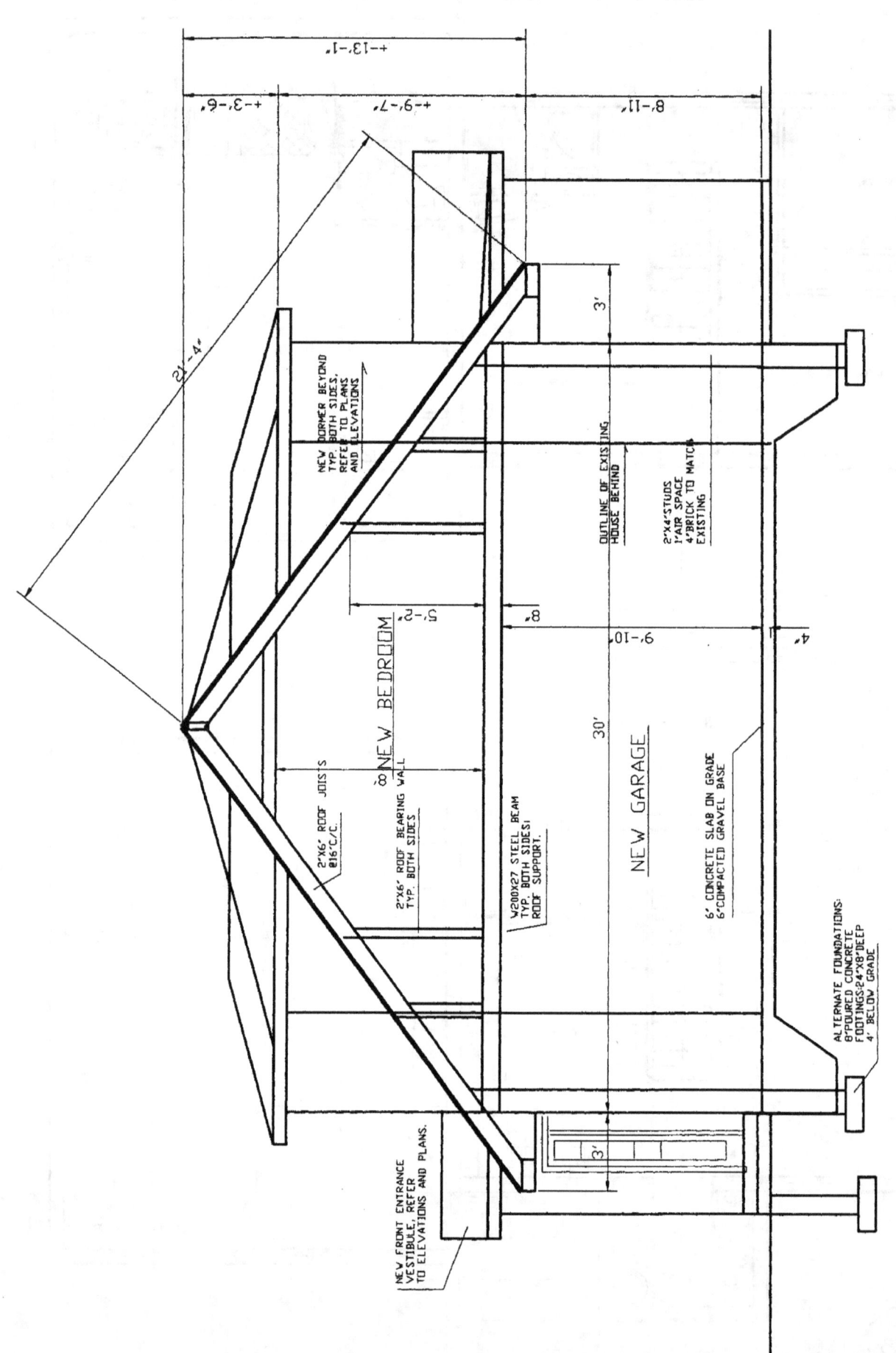

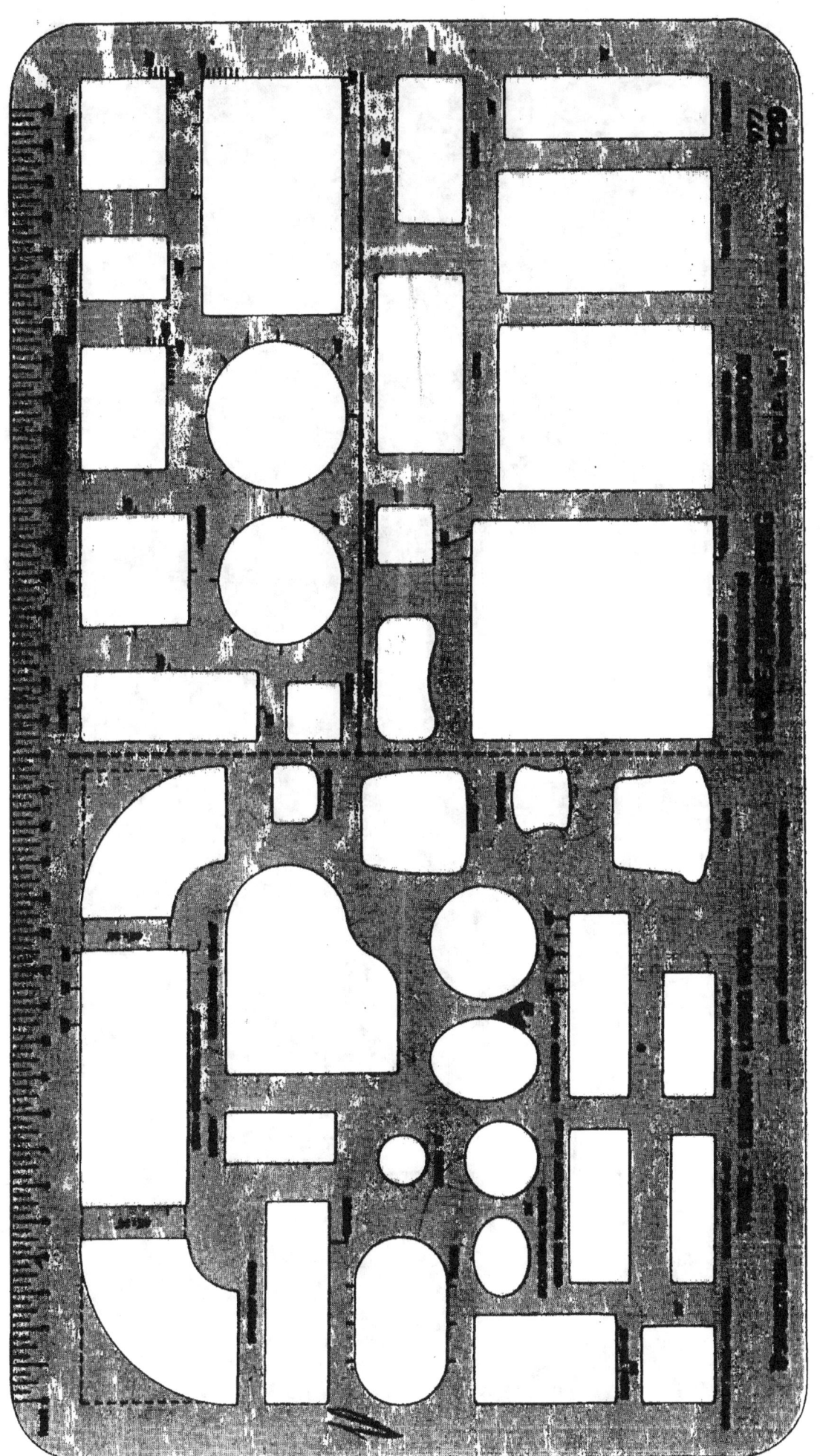

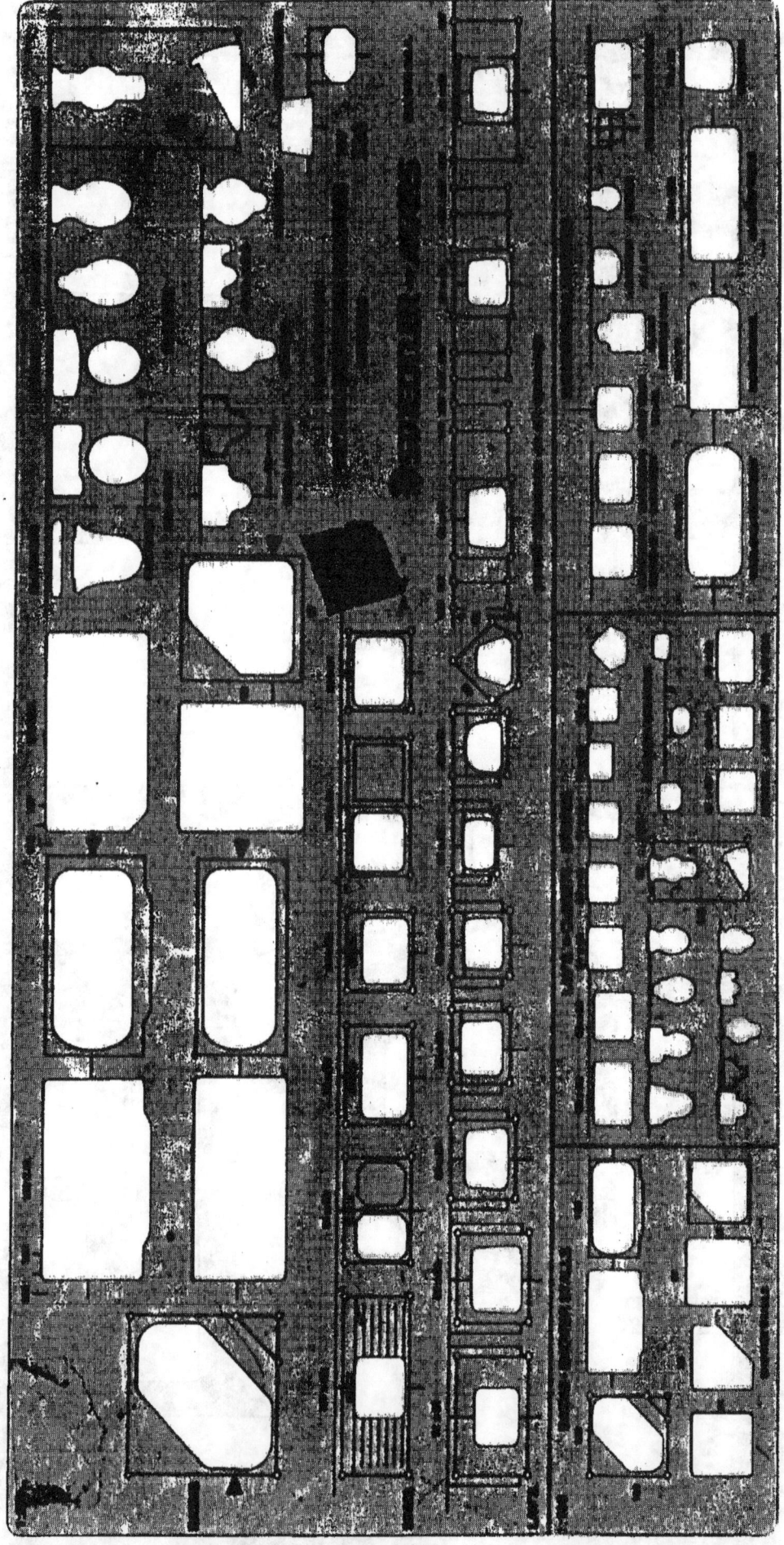

ARCHITECTURAL CONSTRUCTION DETAILS

The handout sheets M1-M4 represent typical masonry veneer and concrete masonry backing construction details.

It is usually a drafter's job to decide on detail scale, sheet size and layout. You will be given this information for the last time this semester. You will have to decide on best scales and design your own layout for the following assignments!

Draw all four details on A2 size sheet. Use 1:5 metric scale. Layout your sheet first by dividing it into four detail boxes. Decide which detail will be drawn in each box. You may want to line up similar details above one another so that wall section components line up. You will draw the same componenents along one construction line.

Pay special attention to quality and precision of your lines. Start with light construction lines and locate all crucial wall detail components. Then show detail elements in section using darkest lines; for elements in elevation view, centerlines, dimensions and notes use a medium line weight. All hatching, construction and guidelines must be the lightest. Use min. 3 line weights.

Construction material sizes are standard if not noted. Design your own titleblock per Mohawk College standards. Your instructor will explain the technical content of each detail. Make sure you understand what you are drafting as these drawings should become part of your portfolio to be used for a job search when you graduate.

You may want to try to complete this drawing on better, but more expensive, mylar paper. Many architectural offices use plastic film instead of inexpensive vellum. Mylar is a stronger, plastic film that stores better and is not damaged by sprinkler water released in case of fire.
For plastic mylar paper you must use E pencil leads available in all professional supply stores. Please note that Mohawk College' bookstore does not carry the mylar paper.

| Mohawk College | BRICK VENEER DETAILS | SCALE NTS | DWG |
| HAMILTON ONTARIO | NAME: Dorota M. Goede | DATE Dec.01' | INSTRUCTIONS |

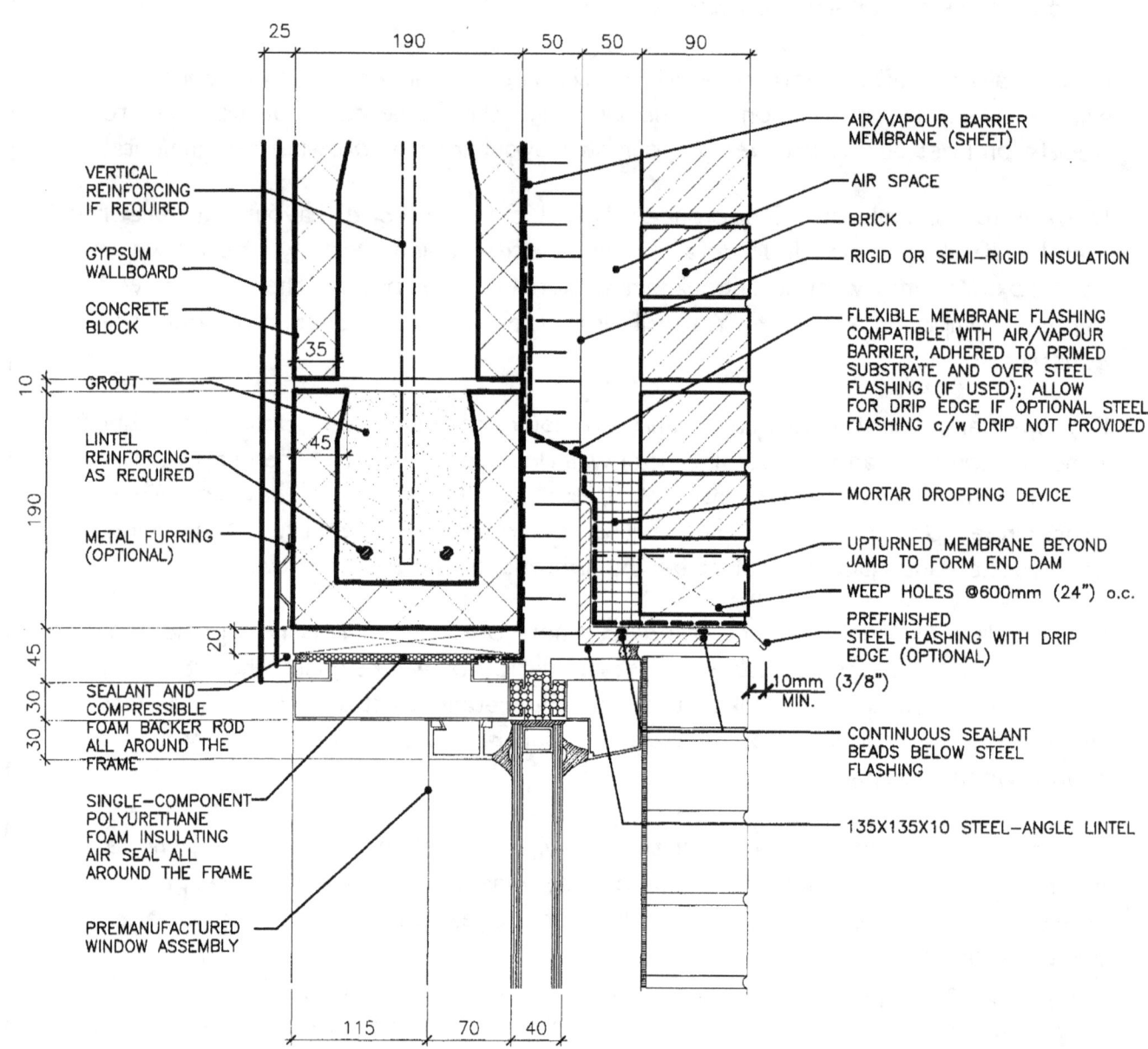

WINDOW HEAD DETAIL

BRICK VENEER DETAILS

Mohawk College — HAMILTON, ONTARIO

D.M.G. Based on CMHC Masonry Canada Details

CLASS: NTS
DATE: Jan. 00
DWG M1

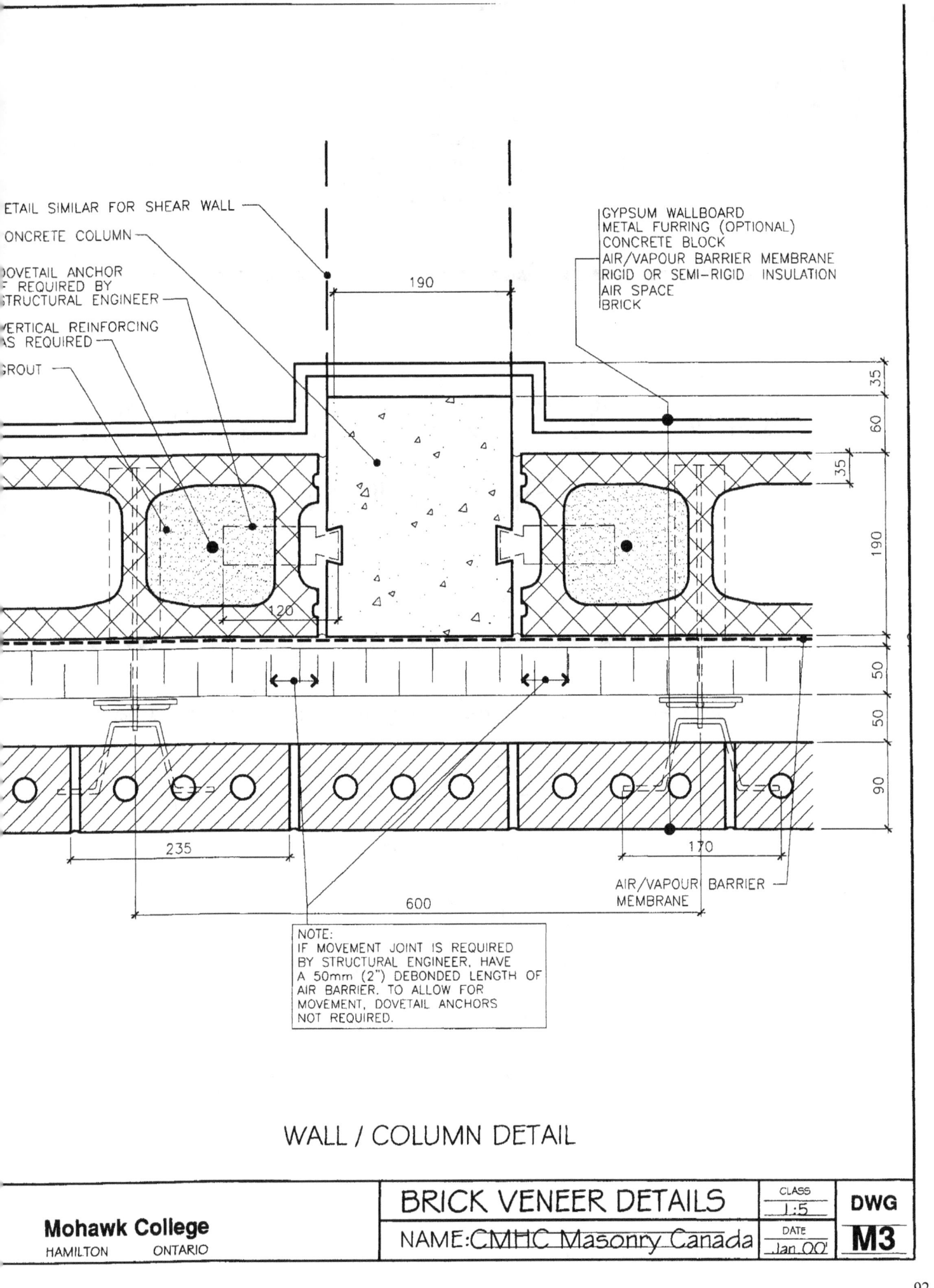

Department of
BUILDING AND CONSTRUCTION SCIENCES
Mohawk College

CIVIL & CONSTRUCTION SECTION

Your Instructor may choose assignments from the following

OBJECTIVE MODULES:

ADVANCED GEOMETRY

FOUNDATION DETAILS

FOOTING REINFORCEMENT

STRUCTURAL PLANS AND SECTIONS

SURVEYING PROFILES AND CALCULATIONS

PARABOLAS & ELLIPSES

Applied geometric construction is one of the basic training areas in drawing. A solid foundation in the knowledge of constructions provides the architectural-civil-transportation student with the methods for drawing general graphic shapes, the use of the bow compass, and the intersection of various lines.

1. Figures below show the basic construction of a parabola within a rectangle. The Instructor will also show you how to divide vertical and horizontal lines into equal parts.

2. Construct an inverted parabola using the rectangle method. The base shall be 2X8260 mm and the height shall be 6840 mm. Divide the appropriate lines into four or five equal parts.

3. Figures below also show the basic construction of an ellipse using the parallelogram method. The construction lines are shown for the upper left quadrant only in the figures. The student will be again required to divide lines into equal parts.

4. Construct an ellipse using the parallelogram method. The long axis shall be 16250 mm and the shorter axis shall be 6840 mm. (same as parabola). Divide the appropriate lines into four or five equal parts.

5. Using an A2 sheet, do both constructions at a scale of 1:50. Leave all construction lines on the drawing. Use standard Department Title Block as shown in the Appendix. Dimension both drawings and place appropriate titles beneath each.

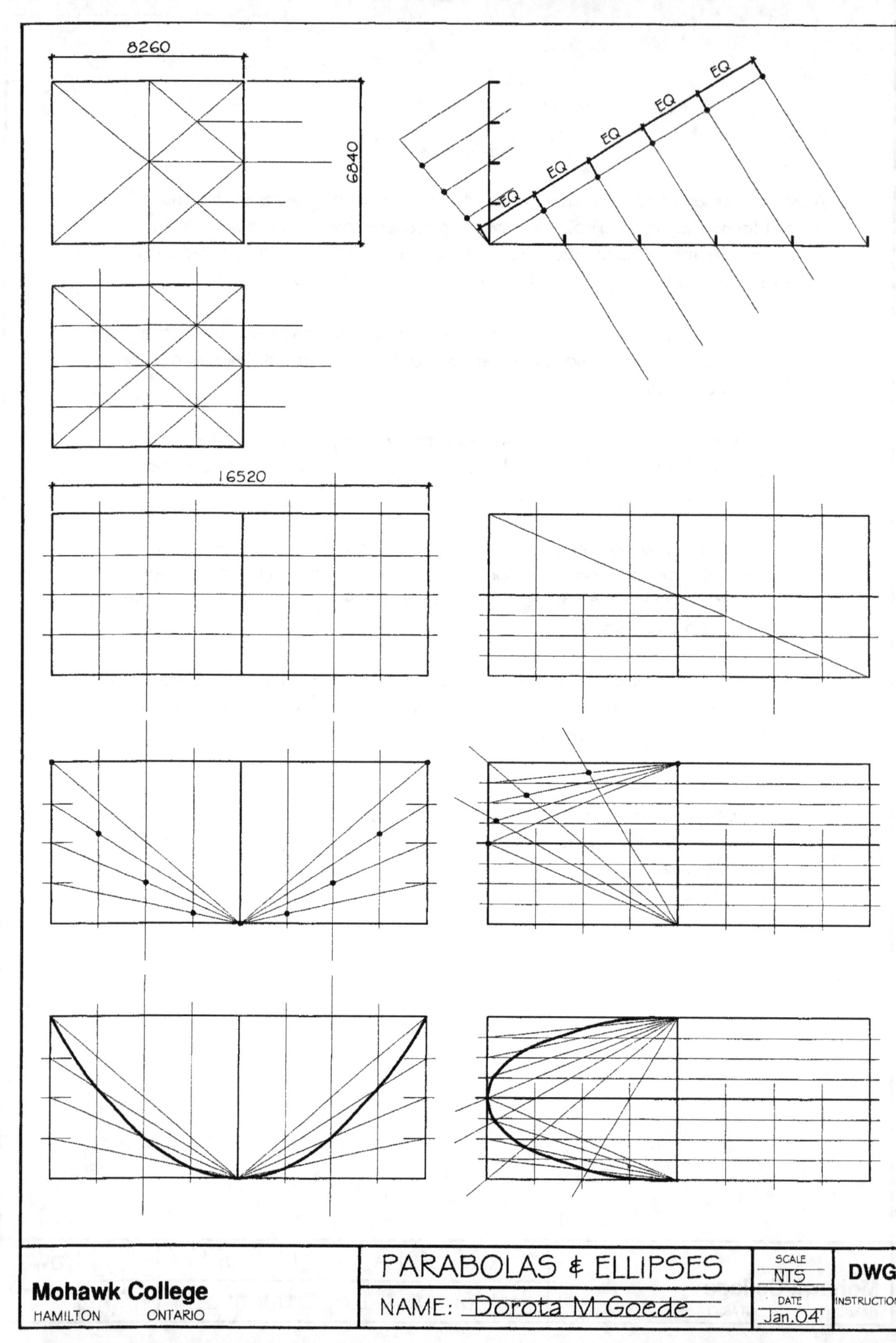

PARABOLAS & ELLIPSES
Mohawk College — HAMILTON ONTARIO
NAME: Dorota M. Goede
SCALE: NTS
DATE: Jan. 04

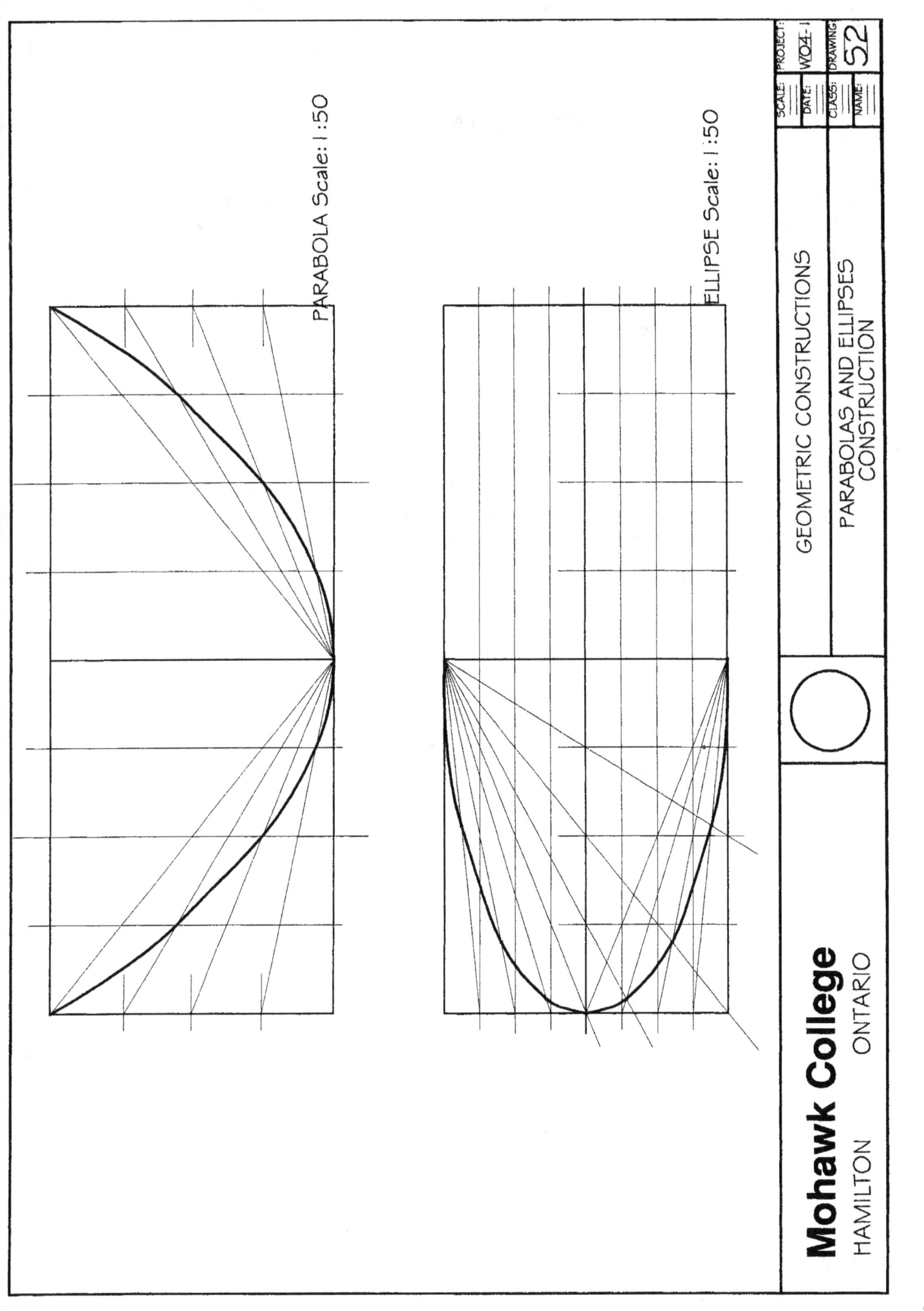

SITE PLAN – FALCON ACRES

Draw the site plan as shown on the following page. Use A2 size sheet. Decide on most suitable scale. Use surveyed property line lengths to create site geometry for each proposed house on lots 5, 6 and 7. Notice that property #6 has a square, 90deg corner. You should start your geometry constructions from that point. Take the page out of your binder and tape it to the board so that the 90deg angle is square to the board. Draw property #6 first. Use both triangles and transfer lines, parallel to those on the handout site, in order to construct the remaining lots 4, 5 and 7. Lot #4 is 70' wide and must fit on your sheet entirely. Also include the entire outline of lot #7.

The Pimlico Drive is 60' wide. Propose houses on the four constructed lots. All lots frontages are 70' unless noted otherwise. Articulate proposed roof lines, show proposed, 20' driveways.
The objective of this exercise is to practice line precision, review text placement using 3mm guidelines, as well as to further practice the use of set square triangles.
Your drawing should represent your preliminary Site Plan proposal for a new subdivision project.

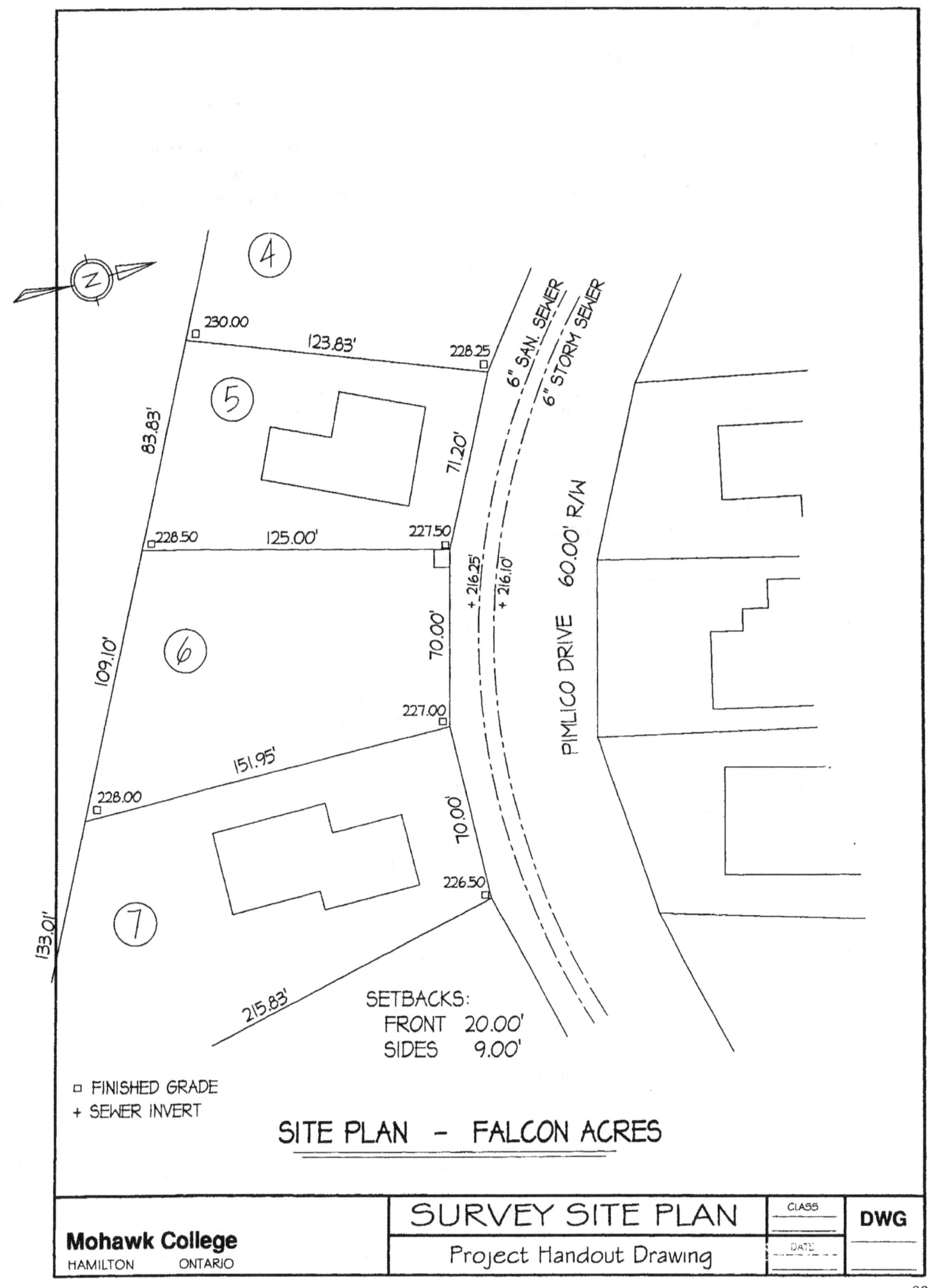

STEEL FRAME ROOF PLAN
BASIC LINEWORK AND PRINTING REVIEW DRAWING

Your Instructor will start with this drawing in order to assess your drafting abilities at the beginning of the second semester. The objective of this assignment is to review your linework, printing, dimensioning, use of metric scale, and ability to design a drawing layout.

Draft Steel Roof Plan as shown. The 3D isometric view will help you visualize the construction details. Design your sheet layout carefully at the beginning of this project. The handout on the next page is drawn in 1:200 scale. Your drawing will be done on A2 size vellum sheet in 1:100 scale. Use light construction lines and outline standard Mohawk College titleblock, then locate the sheet center. Use your scale and establish construction lines that will start and end all grid centerlines. Allow enough space for dimension lines and all notes as shown. Draw two construction lines 12mm apart. This is where you will draw, 12mm, grid circles. The overall rectangle consisting of all required information to be drafted later should be laid out centrally on your sheet. Do not erase any construction or guide lines.

Draw four horizontal centerlines using medium weight lines. Start and end these lines at the precise locations already established. Draw vertical centerlines next making sure that your dash-dot lines intersect. Guide your hand and create the line breaks on the outside of the intersections. You will need the precise intersection points in order to draw the columns later.

Columns are outlined next. Draw light and continuous construction lines boxing out each column. Columns are usually represented as symbols as they are too difficult to draw to scale; the dimensions are simply too small in 1:100 scale to draw them precisely. Draw, therefore, three sets of exactly the same boxes about 400X400mm each. You are ready now to draw the actual columns. Press your pencil quite hard and draw dark and thick, but precise, columns. They should be all the same and located exactly at the intersections of the centerlines.

Draw 400mm concrete block wall next. Use dark lines and construct all corners precisely. Draw wall hatching very lightly; lines must be on 45deg angle approximately 3mm apart.

Grid centerlines are labeled with 12mm circles. Use medium pencil weight. The grid notations can be slightly larger, 4-5mm.

Complete your drawing by adding all dimensions and notes as shown. Use 3mm construction lines for all text; the lettering guide can help you achieve much better printing results. Use the vertical edge against the tee-square for all vertical elements of letters and numbers. Finish your titleblock using about 5mm size text. Store your drawing flat or rolled up, the drafted side on the outside.

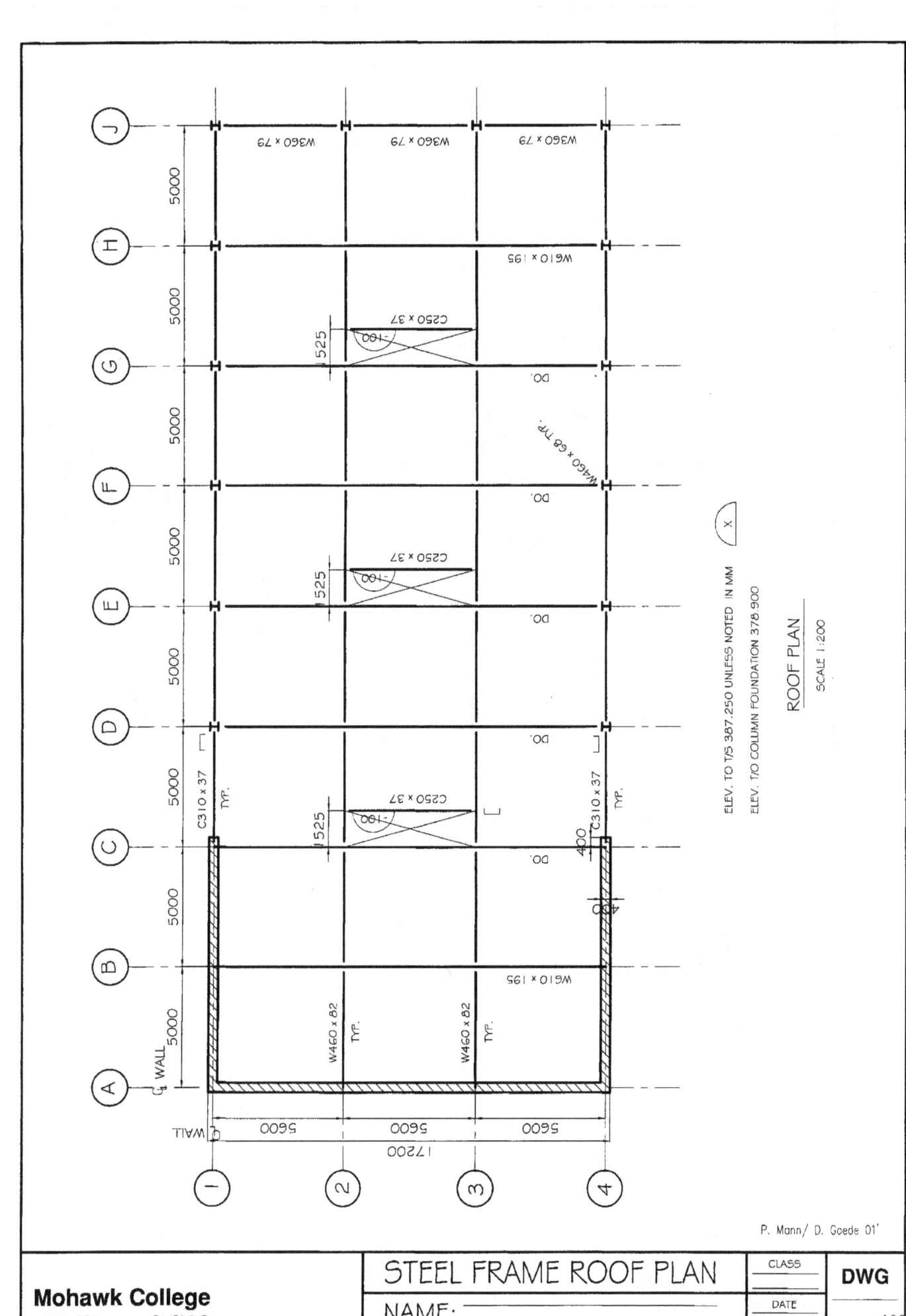

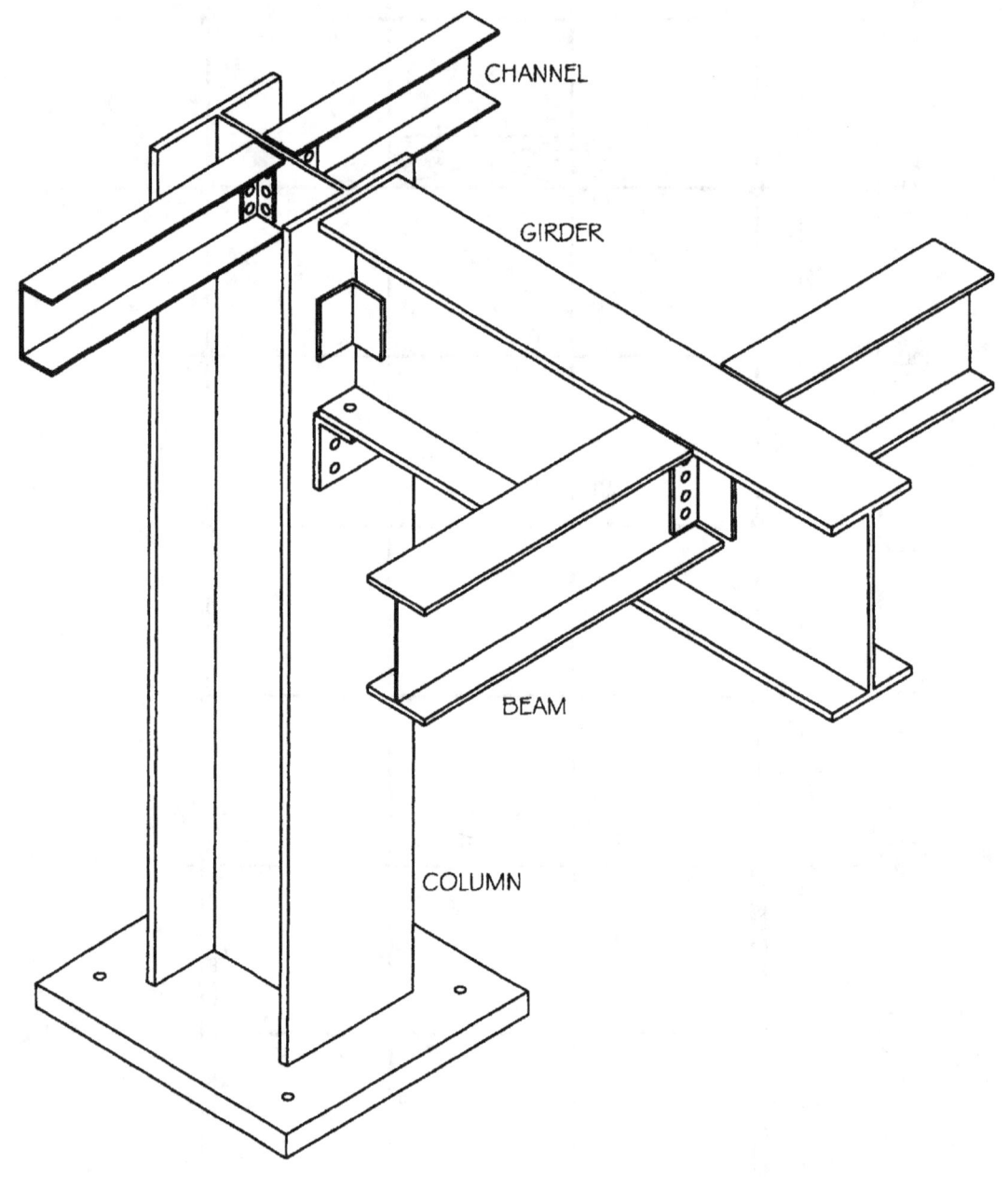

STEEL FRAME ROOF PLAN		
NAME:		DWG

Mohawk College
HAMILTON   ONTARIO

P. Mann Sept.97

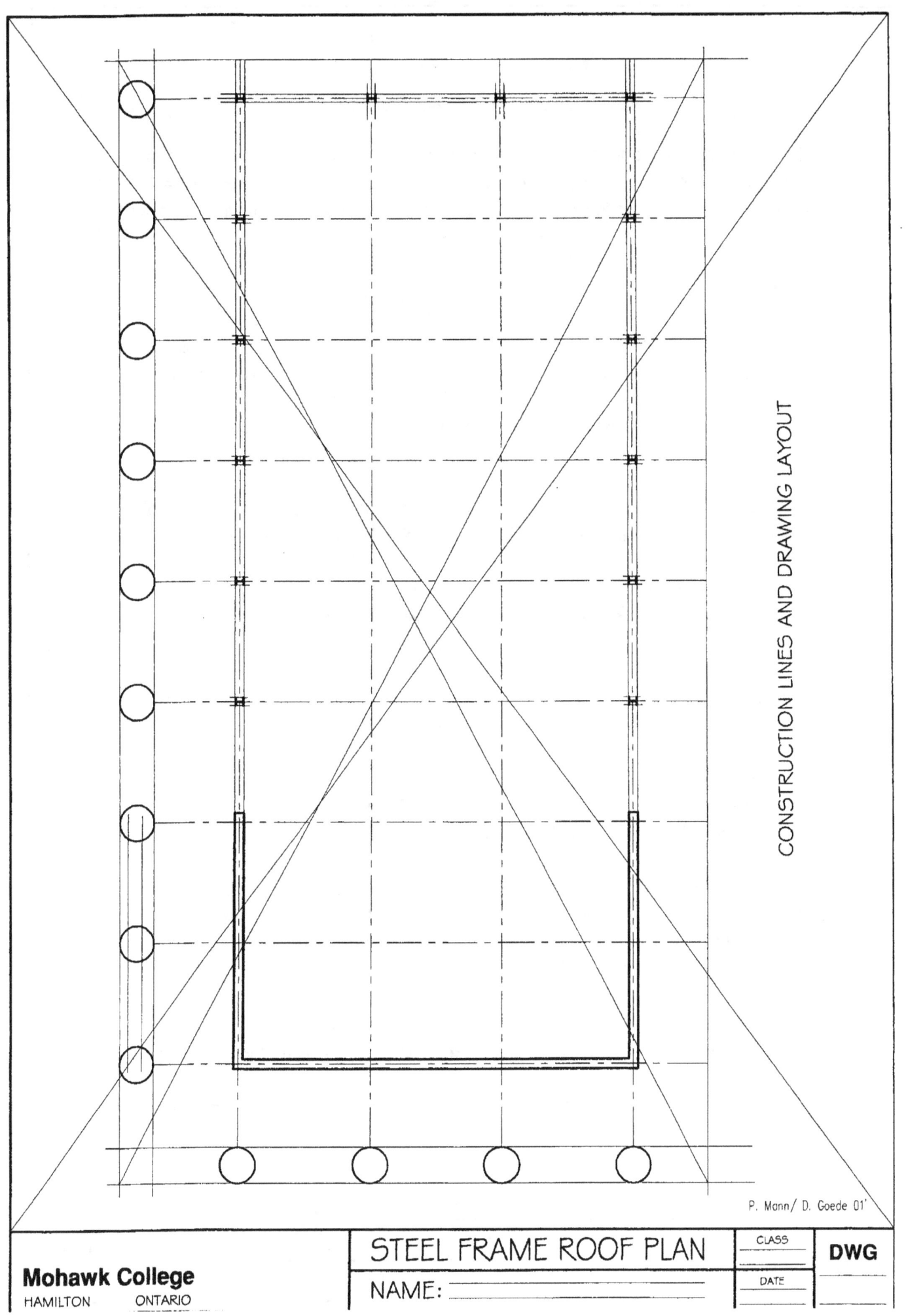

HILL & DALE SUBDIVISION PROFILE

Complete the drawing and calculations as indicated in the instructions below. Use two sheets of A4 size paper in which to do the profile drawing. Read instructions carefully before starting, to ensure your final drawing includes all the required elements and measurements, and is laid out properly.

1. The attached figure shows the proposed alignment of a sewer in the Hill & Dale Subdivision which is being serviced in preparation for residential construction. Your first task is to determine the existing ground elevations at the 10 metre stations along the proposed sewer between Manhole #10 (1+100.000) and Manhole #12 (1+200.000). Interpolate between contours to get the ground elevations in metres above sea level (m. ASL) estimated to the nearest 0.1 metre.

2. Construct a profile drawing. The profile should have a horizontal scale of 1:500 and vertical scale of 1:100, and show the following:

    - original ground line and elevations (m. ASL)
    - proposed sewer profile and invert elevations (m. ASL)
    - chainages using the 10 metre intervals

3. In order to locate the sewer on the drawing use the following sewer invert elevations at the corresponding manholes:

    - Manhole #10   Invert Elevation 97.50 m. ASL
    - Manhole #11   Invert Elevation 96.00 m. ASL
    - Manhole #12   Invert Elevation 94.00 m. ASL

    Make sure you include chainage, sewer invert elevation, and the original ground elevations for every 10 metre station. Also include the datum line elevation and % slope of the sewer between Manholes #10-11 and Manholes #11-12

4. Complete your drawing with a three line title block, borders, and appropriate titles and any notes which you deem necessary to make it understandable.

    P.S. The drawing will fit easily on the sheet if you do a little planning in advance.

Mohawk College
HAMILTON   ONTARIO

HILL & DALE SUBDIVISION
NAME: C. Blackwood

SCALE: NTS
DATE: Dec.01'

DWG
INSTRUCTIONS

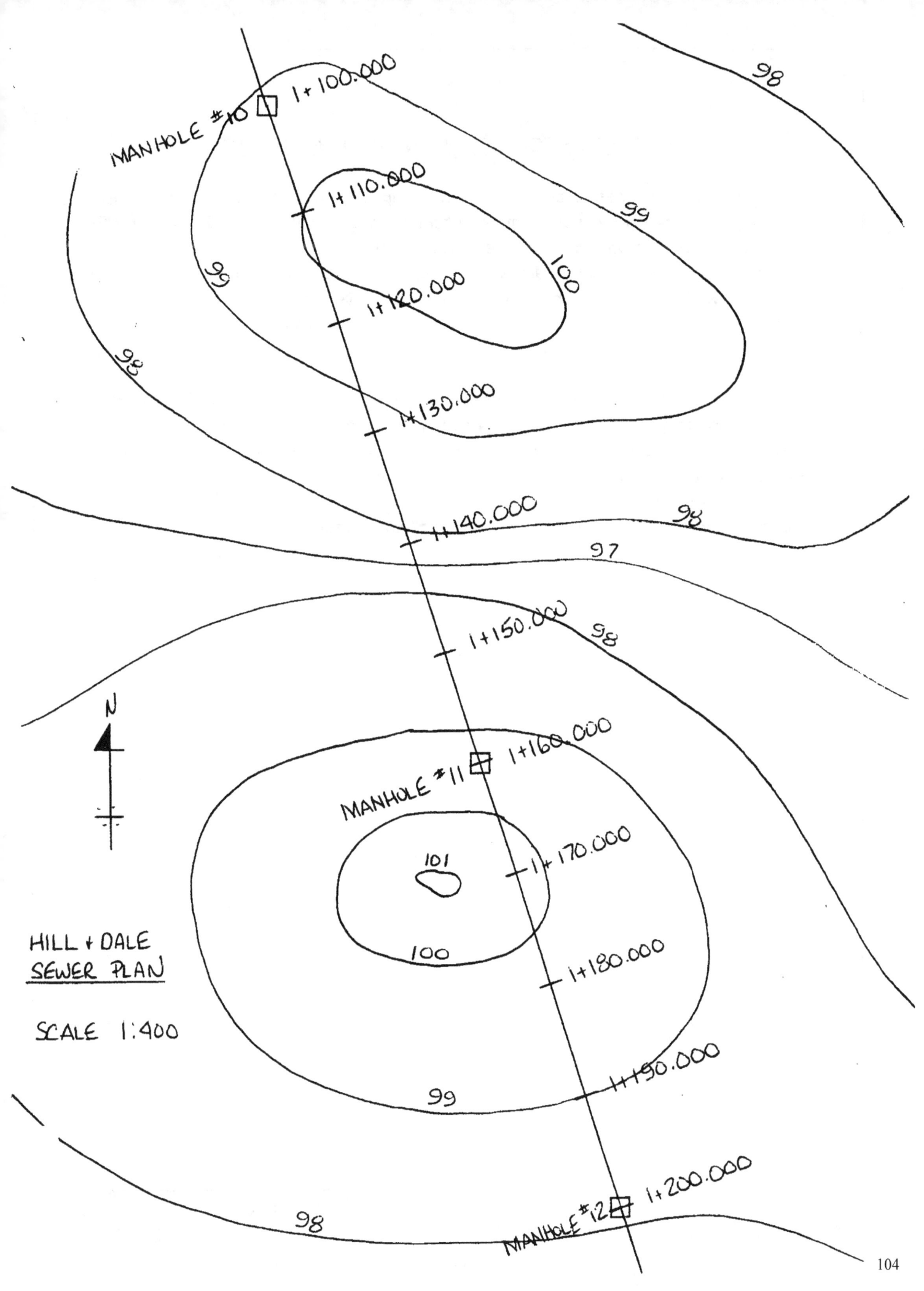

STEEL BEAM CONNECTION AND BEARING DETAILS

The objective of this assignment is to further improve your linework, printing and dimensioning. Draw the steel beam and steel bearing details in 1:10 scale on A2 size sheet as shown on the last page of this module.

Note that the length of the beam is not shown to scale. It is an industry standard not to show the entire length of the beam. This allows for connection details to be represented clearly in larger scale. All steel shop drawings are drafted this way.

Design the layout of your A2 size sheet with light construction lines. Start each detail by drafting the center lines of a structural member. Construct the center lines for all bolts. Use a small circle template for drafting the precise bolt hole circles. Add dimensions and notes. Complete the drawing with concrete and steel hatching.

Frame your work with a standard Mohawk College titleblock.

Mohawk College
HAMILTON ONTARIO

BEACH AUTO SHOP

NAME:

CLASS

DATE

DWG

W-SHAPES

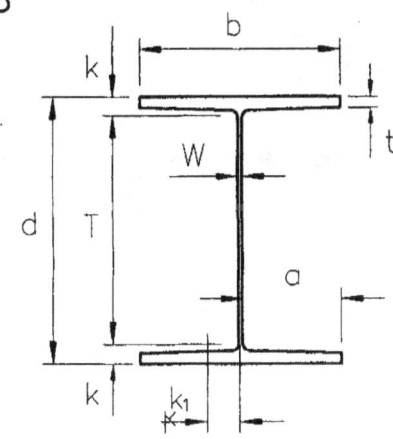

SIZE	d	b	t	w	a	T	k	k_1
W610x195	622	327	24	15	156	535	43	25
W460x82	460	191	16	10	91	394	33	21
W460x68	459	154	15	9	73	396	32	20
W360x79	354	205	17	9	98	281	36	22

C-SHAPES

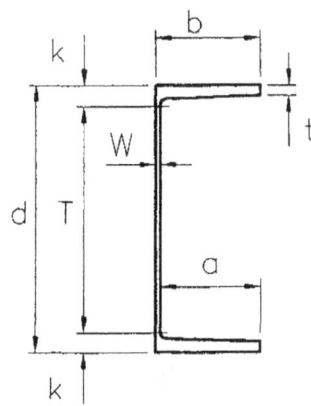

SIZE	d	b	t	w	a	T	k
C310x37	305	77	13	10	67	250	28
C250x37	254	73	11	13	60	206	24

P. Mann Sept 97

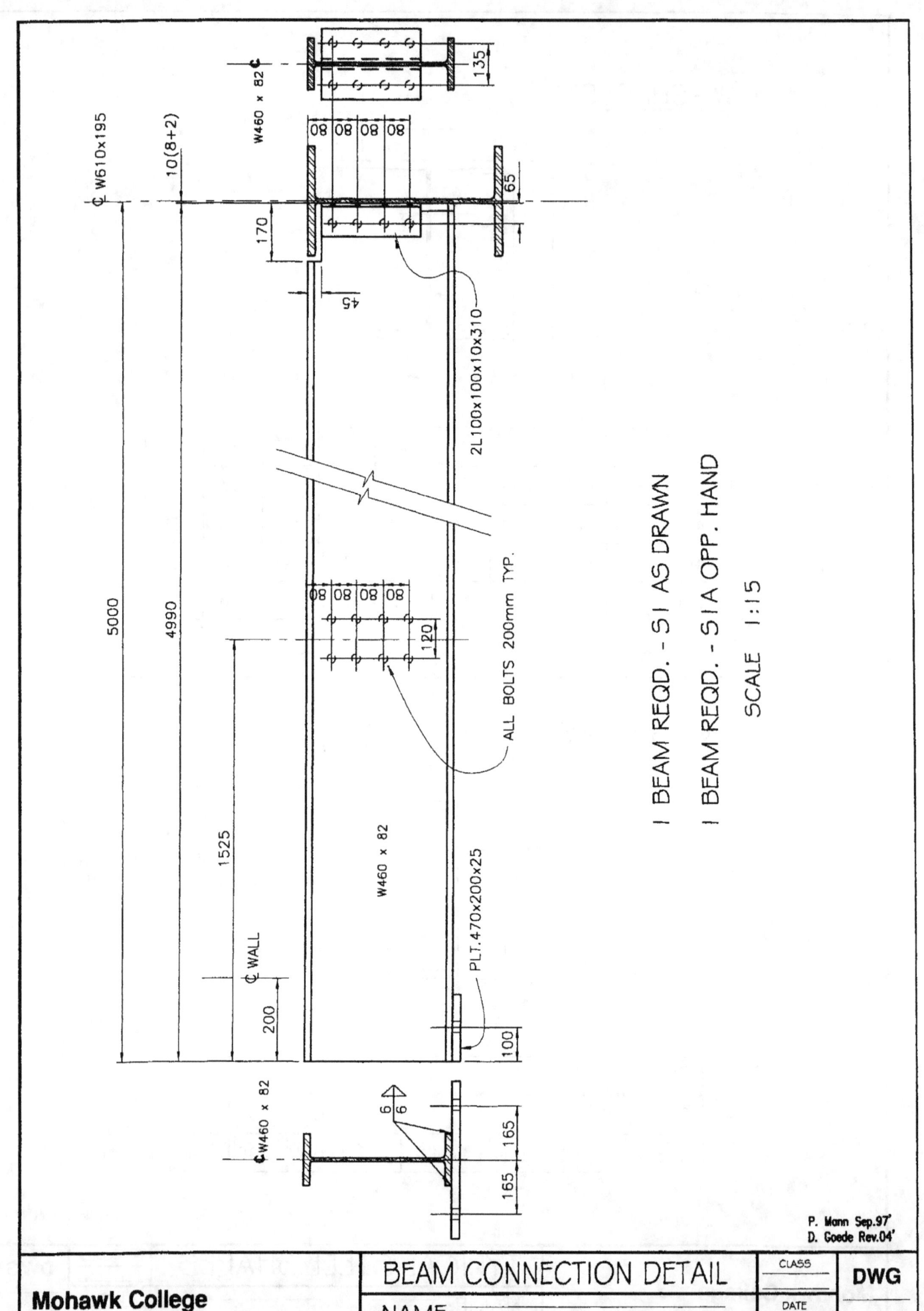

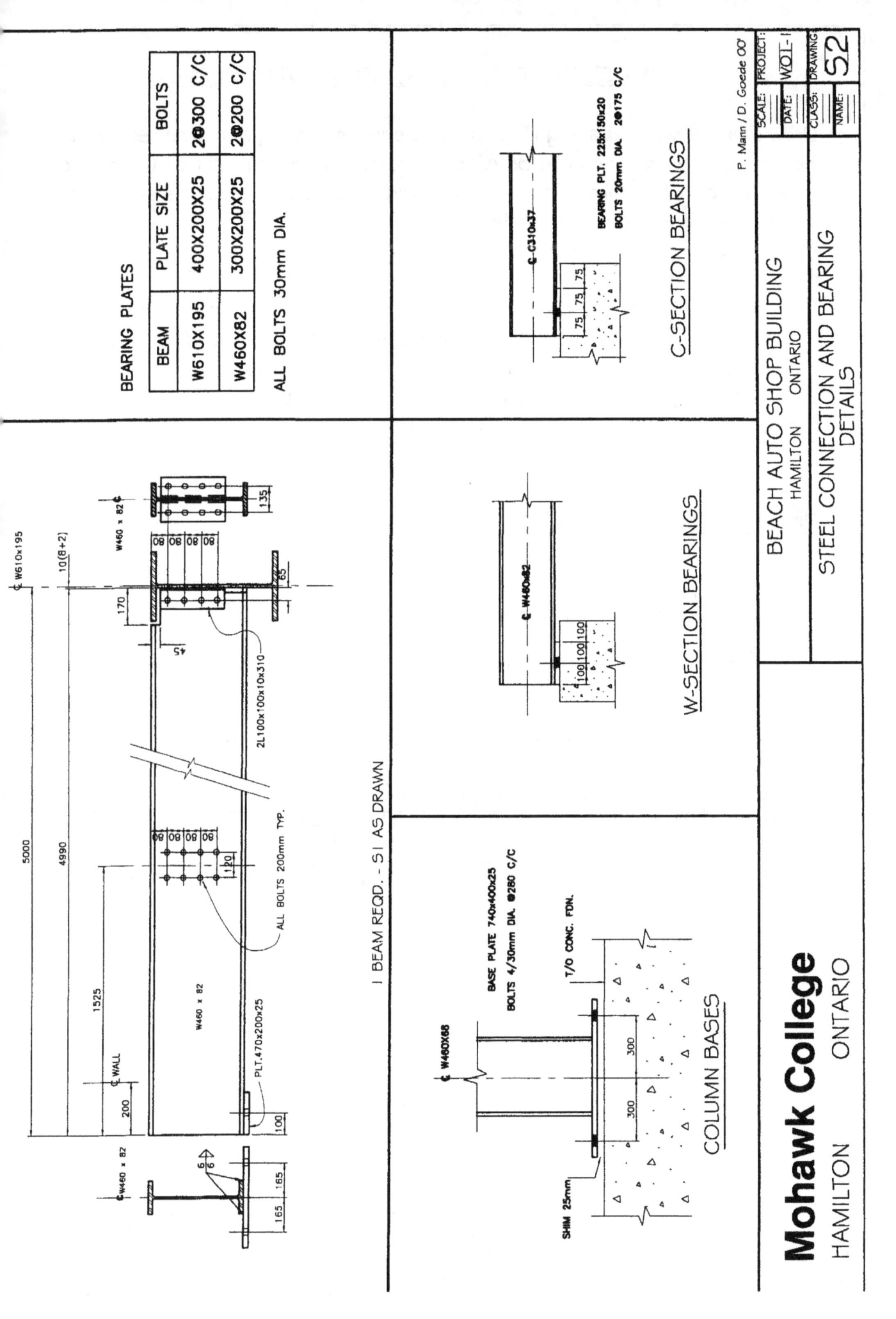

ARMTREE INC. PROJECT - REINFORCEMENT DETAILS

Draw Plan layout on drawing R1 and three sets of footing reinforcement details from pages R2, R3, and R4, on A2 size sheets. Propose layout and decide on most appropriate scales. Treat the attached handout as engineer's sketches. They are drafted but not to scale!
Do not trace the handout.

Draw reinforcement bars the darkest as they are the main content, the object, of this project. Use a medium line for all descriptive information and dimensions. Use light construction lines and guidelines. The ground line must be drafted dark as it is in section; ground hatching must be shown very light.
Do not erase any very light construction lines.

Add dimensions and notes. Frame your work with standard Mohawk College titleblock.

NOTES: COVER 50 mm
FOR LENGTH OF BARS ALLOW FOR COVER
FOR NO. OF BARS OMIT COVER
USE "0. 3 ROUNDING RULE"
  (E.G., 7. 3 SPACES = 8 SPACES = 9 BARS)
FORMS: FOR "B" - ASSUME SLOPING SIDES = HEIGHT

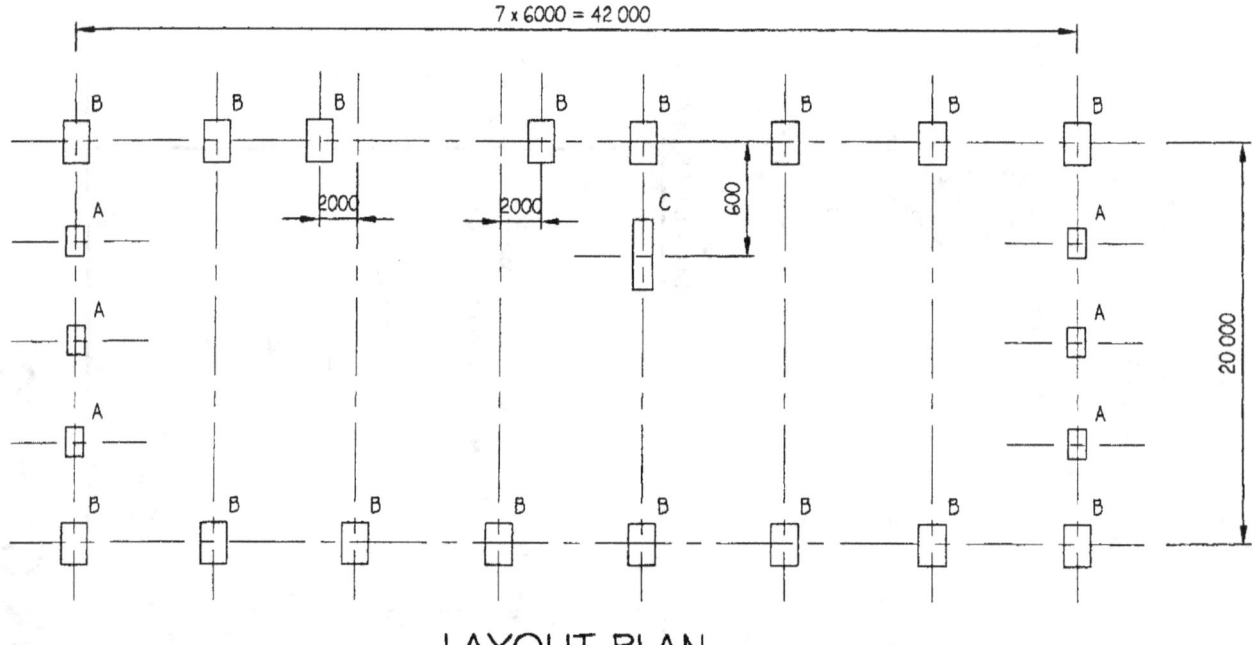

LAYOUT PLAN

P. Mann Sept 97

| Mohawk College HAMILTON ONTARIO | REINFORCEMENT DETAILS ARMTREE PROJECT | SCALE NTS / DATE Mar 01' | DWG R1 |

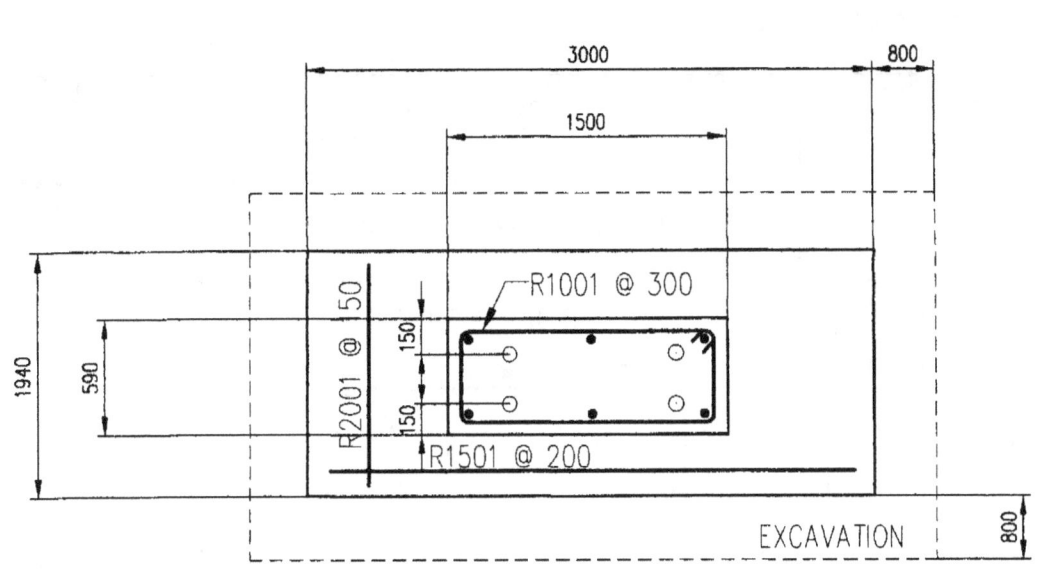

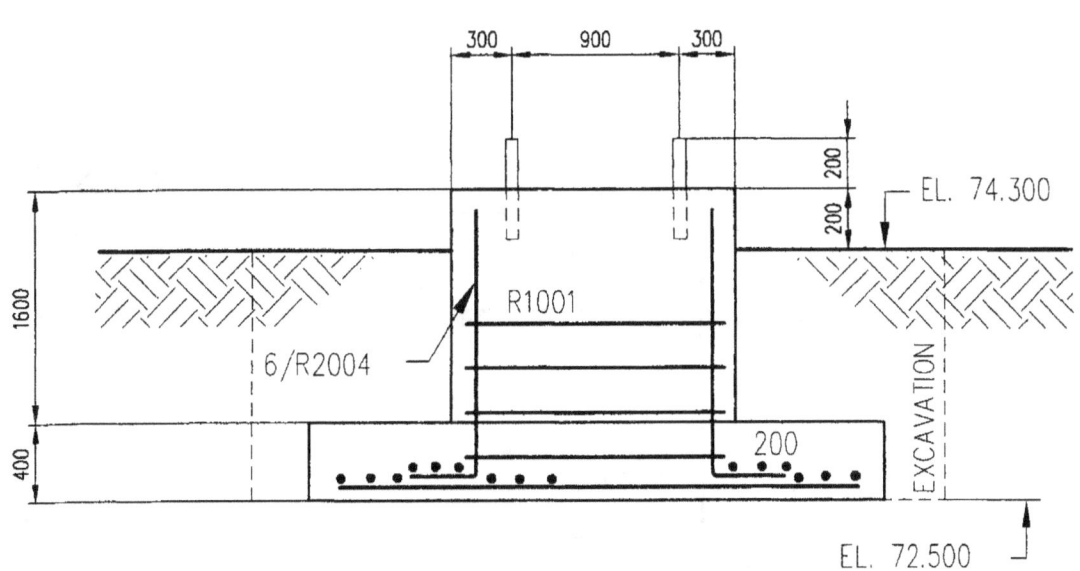

FOOTING TYPE "A" - 6 REQD.

P. Mann Sept 97

Mohawk College HAMILTON ONTARIO	REINFORCEMENT DETAILS	SKETCH NTS	DWG
	ARMTREE PROJECT	DATE Mar 01'	R2

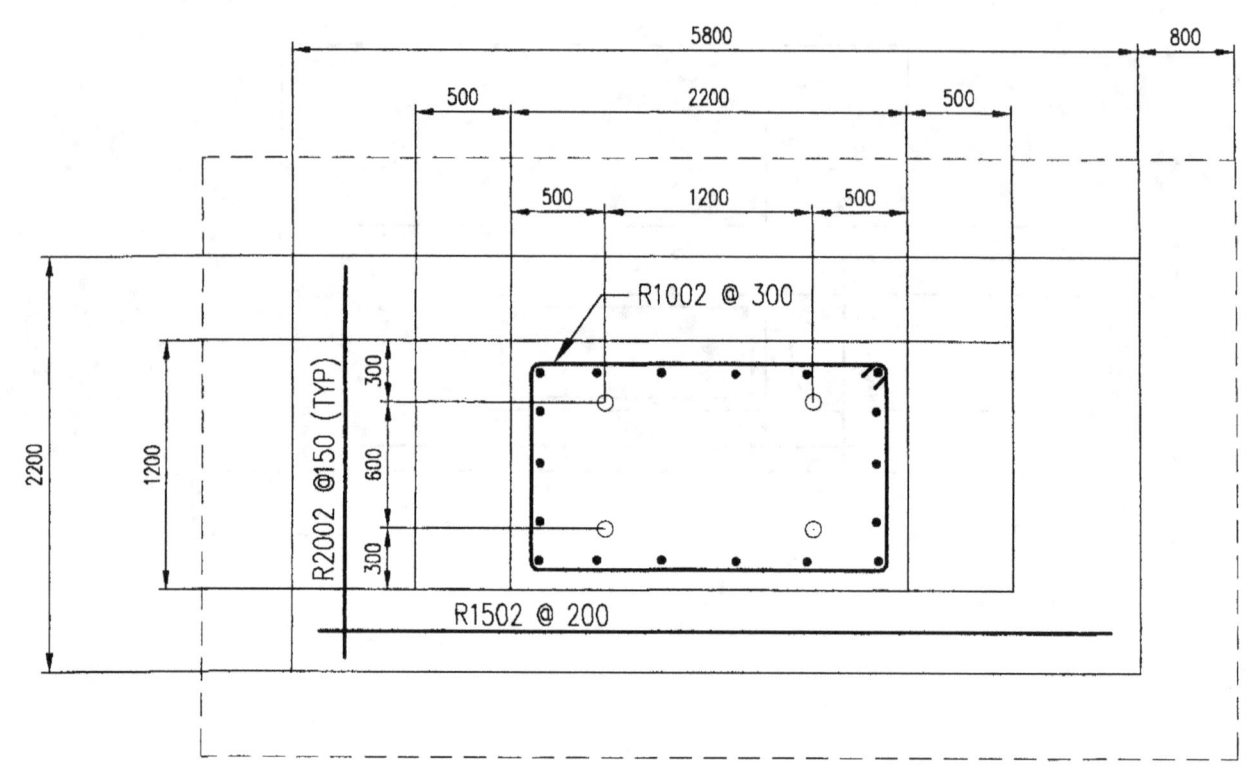
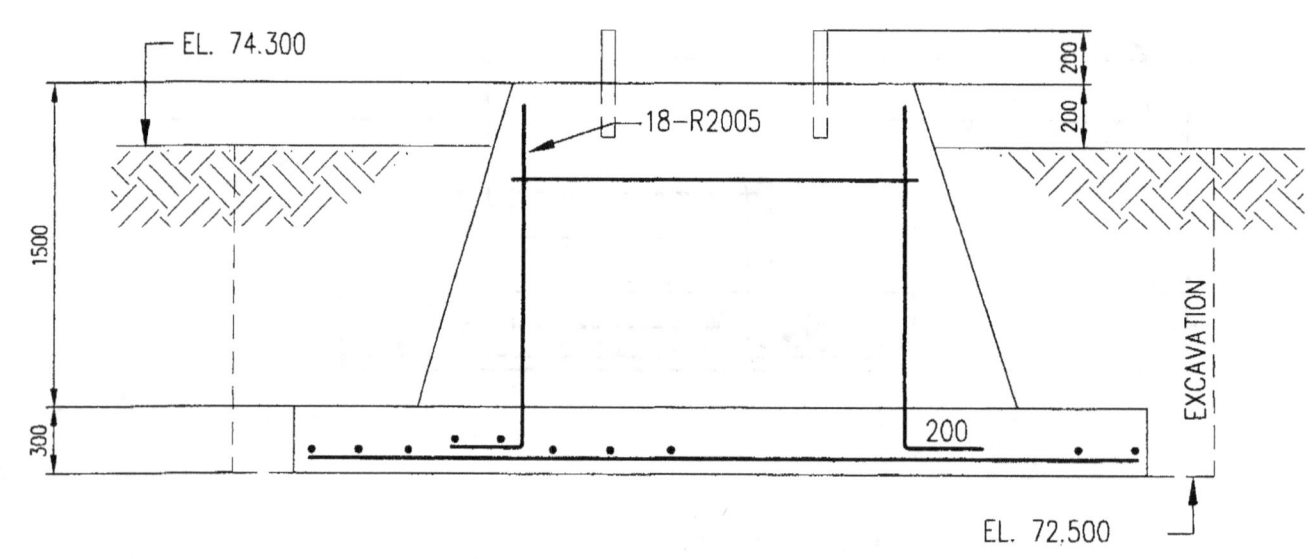

FOOTING TYPE "B" - 16 REQD.

P. Mann Sept 97

Mohawk College
HAMILTON   ONTARIO

REINFORCEMENT DETAILS
ARMTREE PROJECT

SKETCH / NTS
DATE / Mar 01'

DWG **R3**

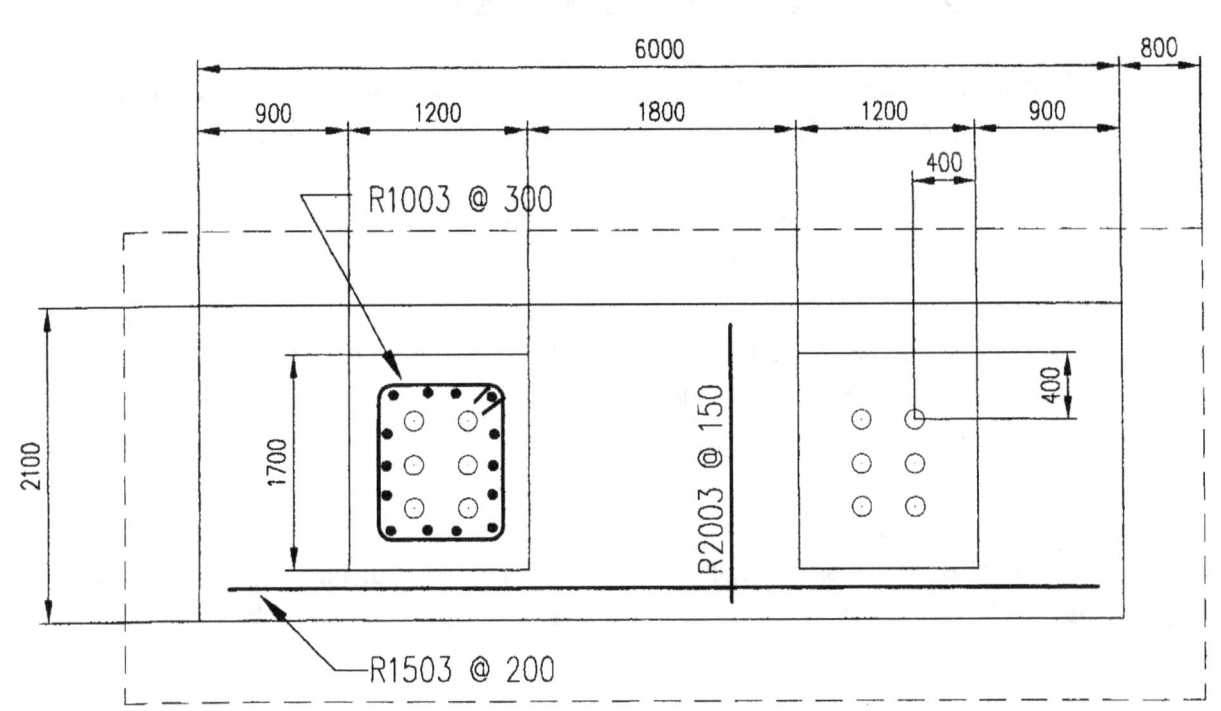

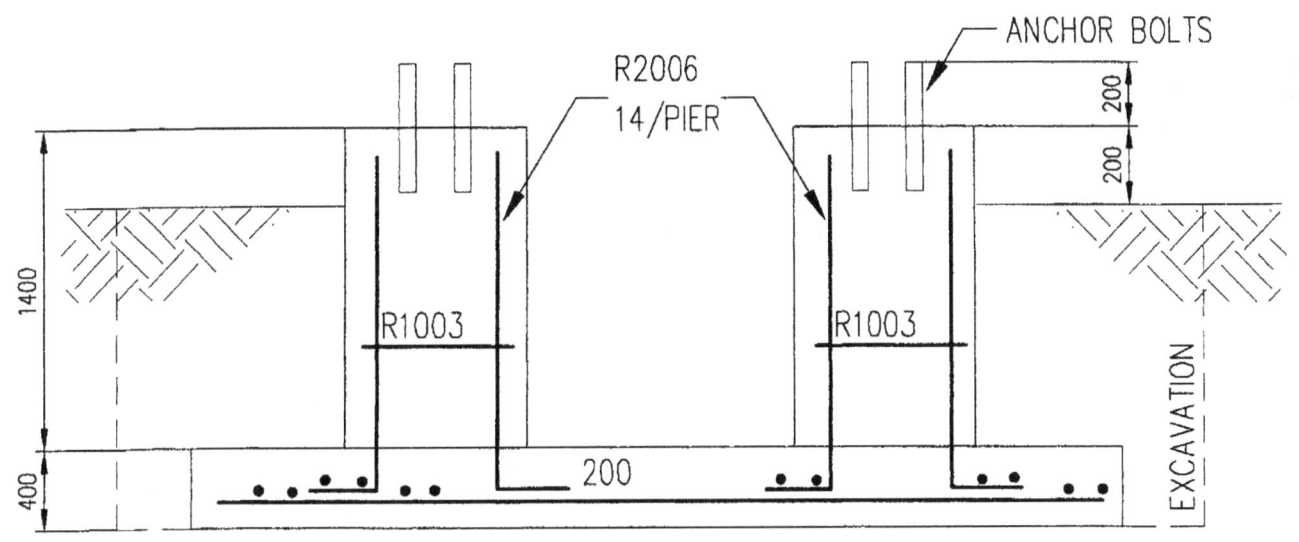

FOOTING TYPE "C" - 1 REQD.

P. Mann Sept 97

| Mohawk College | REINFORCEMENT DETAILS | SKETCH NTS | DWG |
| HAMILTON ONTARIO | ARMTREE PROJECT | DATE Mar 01' | R4 |

CUSTOM STEEL TRUSS CONSTRUCTION DETAILS

The objective of this assignment is to draw a custom prefabrication shop drawing for a steel truss. We will further improve your linework, printing and dimensioning.

Use A2 size sheet for each truss or draw both on A1 size sheet. Design the layout of your sheet with light construction lines. Draw titleblok.

Both trusses shown on attached designer's sketches are constructed from precut to size steel angles. The connections are made via prefabricated in shop steel connection plates.

Start by drafting the centerlines of all structural members. Construct the centerlines for all bolts. Use a small circle template for drafting the precise bolt hole circles. Add dimensions and notes. Complete the drawing with

---

Mohawk College
HAMILTON   ONTARIO

CUSTOM TRUSS DRAWINGS
Instructions

CLASS
DATE
DWG

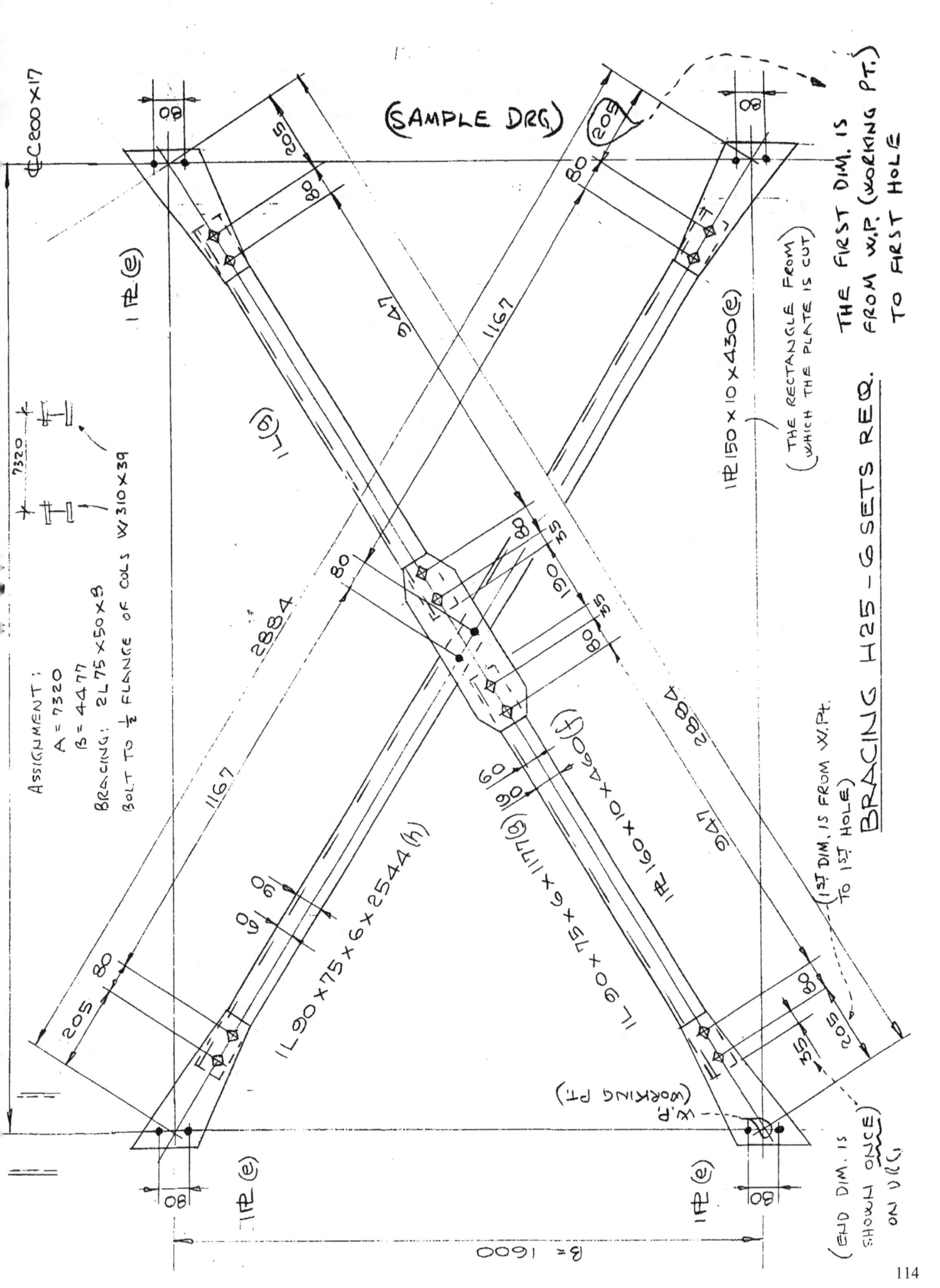

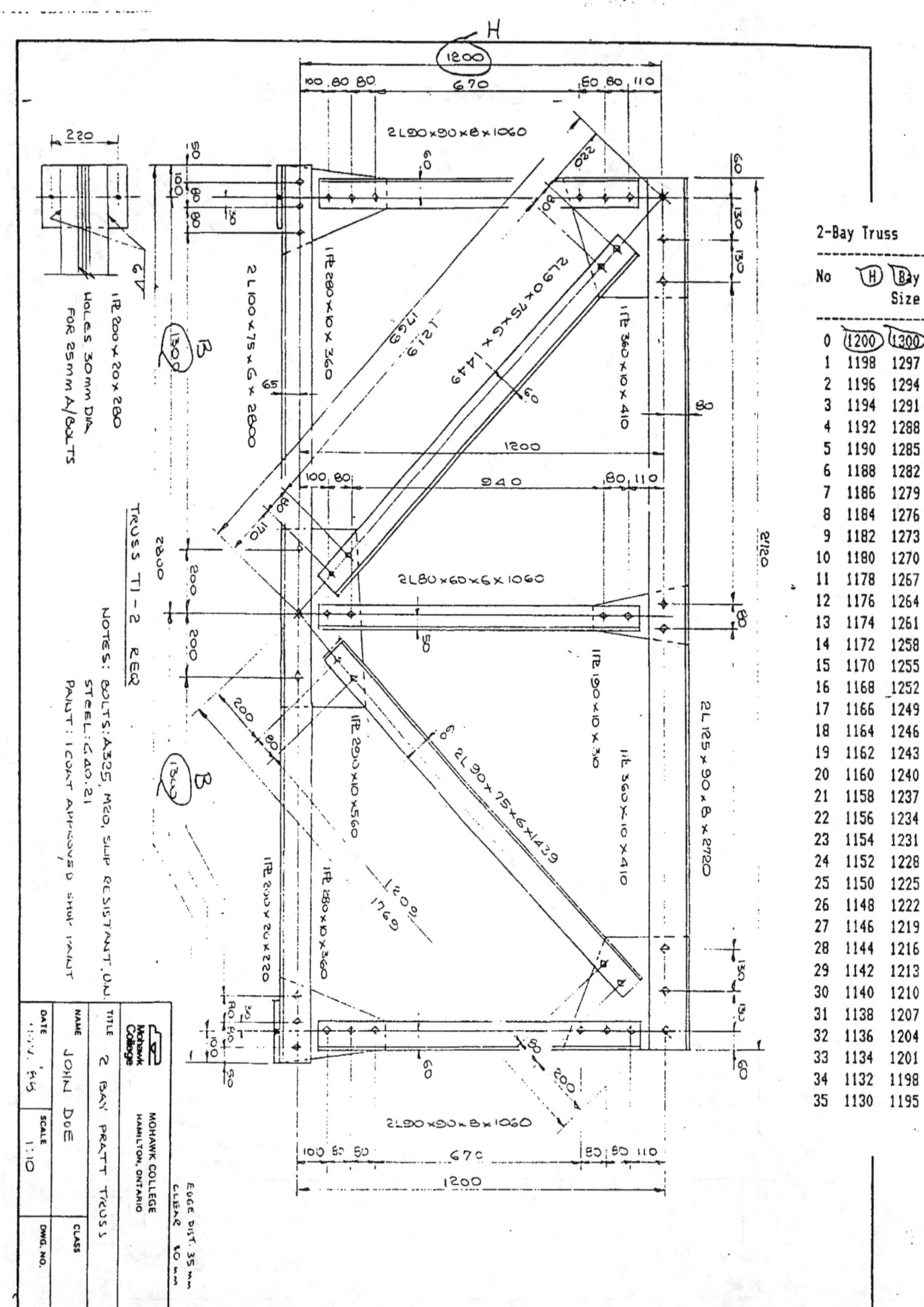

No	H	Bay Size
0	1200	1300
1	1198	1297
2	1196	1294
3	1194	1291
4	1192	1288
5	1190	1285
6	1188	1282
7	1186	1279
8	1184	1276
9	1182	1273
10	1180	1270
11	1178	1267
12	1176	1264
13	1174	1261
14	1172	1258
15	1170	1255
16	1168	1252
17	1166	1249
18	1164	1246
19	1162	1243
20	1160	1240
21	1158	1237
22	1156	1234
23	1154	1231
24	1152	1228
25	1150	1225
26	1148	1222
27	1146	1219
28	1144	1216
29	1142	1213
30	1140	1210
31	1138	1207
32	1136	1204
33	1134	1201
34	1132	1198
35	1130	1195

BUILDING PERIMETER AND FOUNDATION DETAILS

Draw the structural plan and foundation section as shown below. Use an A2 size sheet. Design sheet layout and decide on appropriate scale. Use minimum three line weights. Show all grid and centerlines with dash-dot uniform and consistent lines. Use consistent dashed lines for footings and excavation lines.

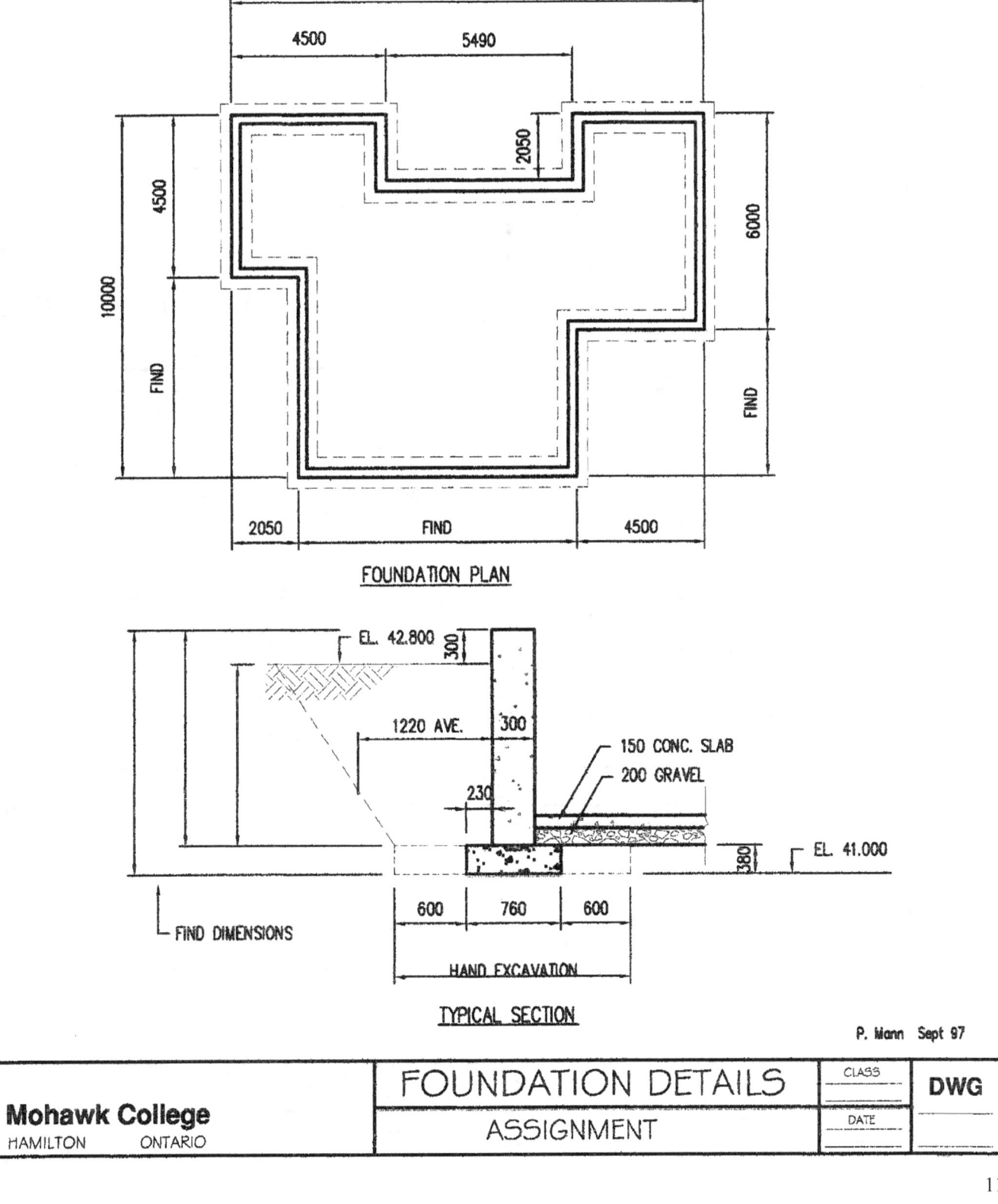

FOOTING REINFORCEMENT DETAILS

DRAW THE DETAILS SHOWN ON THE FOLLOWING PAGES.
USE A2 SIZE SHEET AND 1:25 SCALE.

USE MINIMUM OF THREE LINE WEIGHTS.
NOTE THAT OBJECT LINES HERE ARE THE REINFORCEMENT LINES.

Typical reinforcement concrete cover is minimum 50mm. Draft reinforcement bars as symbols (dark uniform line) and offset from concrete as required.
All Anchor bolts are 20mm dia and 1m long with standard 200mm bend.

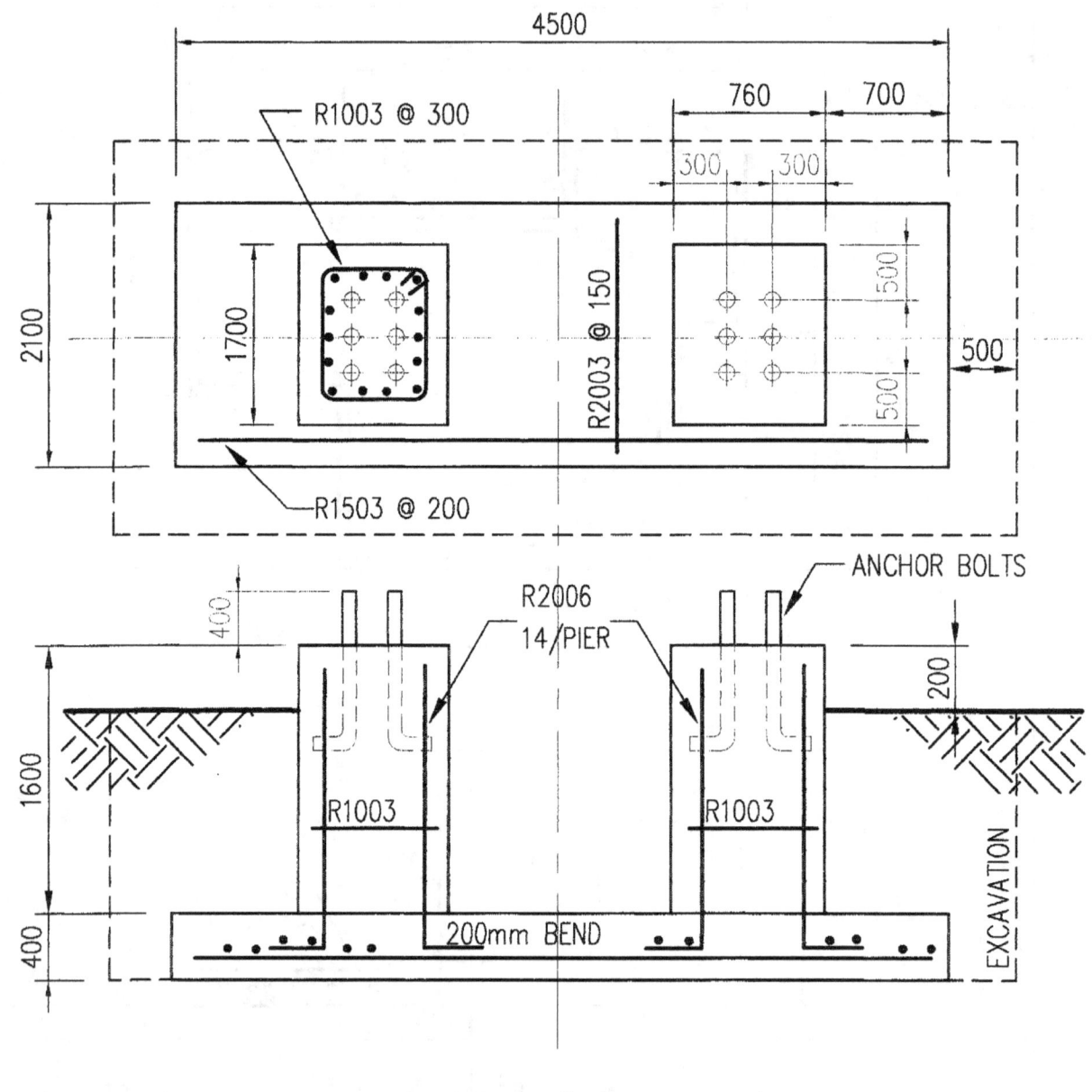

FOOTING TYPE "C" – 1 REQD.

P. Mann Sept 97

Mohawk College — HAMILTON ONTARIO

FOOTING REINFORCEMENT
DRAFTING ASSIGNMENT

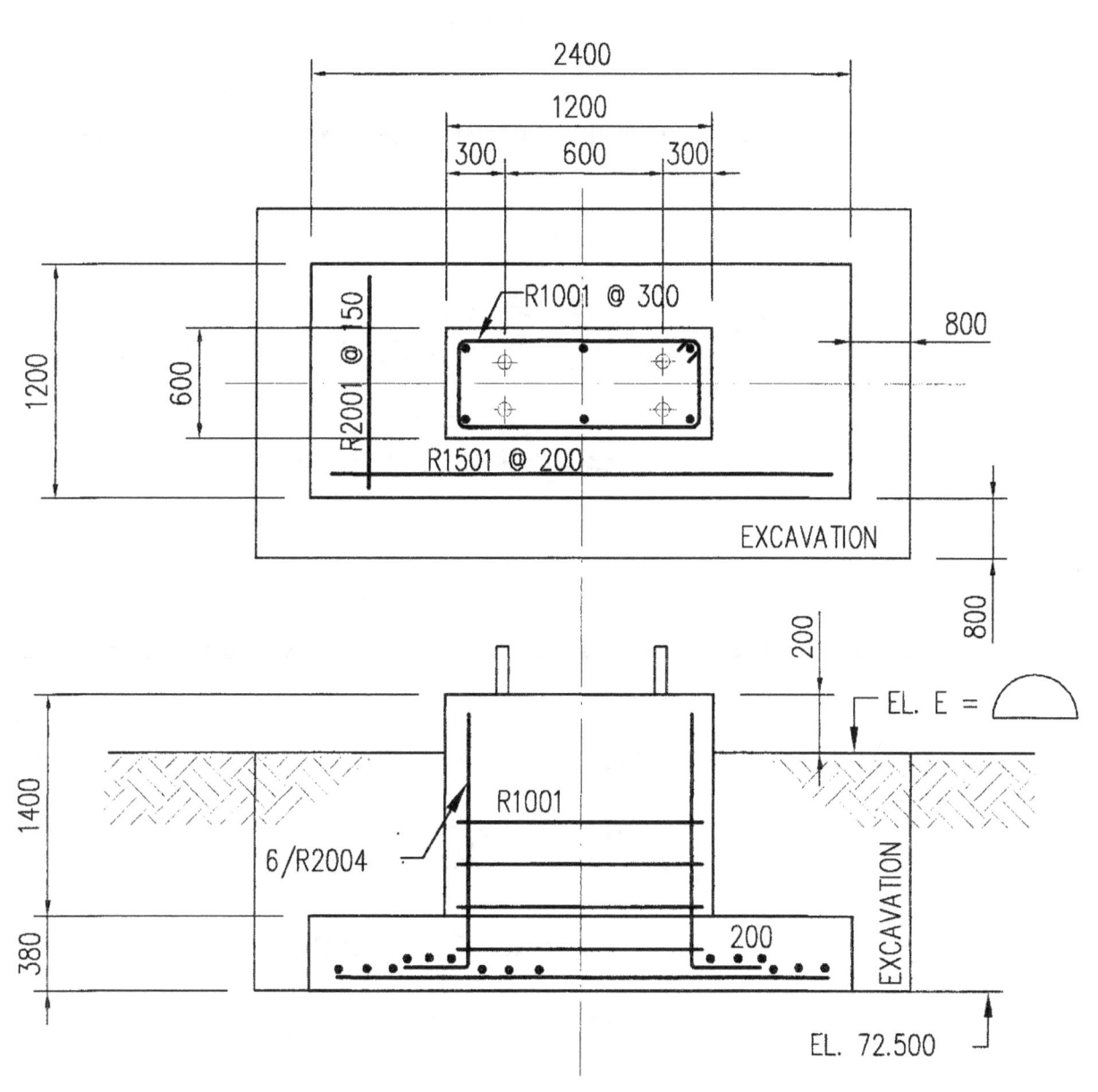

Department of
BUILDING AND CONSTRUCTION SCIENCES
Mohawk College

TRANSPORTATION SECTION

Your Instructor may choose assignments from the following

OBJECTIVE MODULES:

ADVANCED GEOMETRY

ARC CONSTRUCTIONS

INTERSECTIONS

PRELIMINARY TRANSPORTATION DESIGN

R
ASSUMED OR GIVEN RADIUS OF DESIRED CURVE

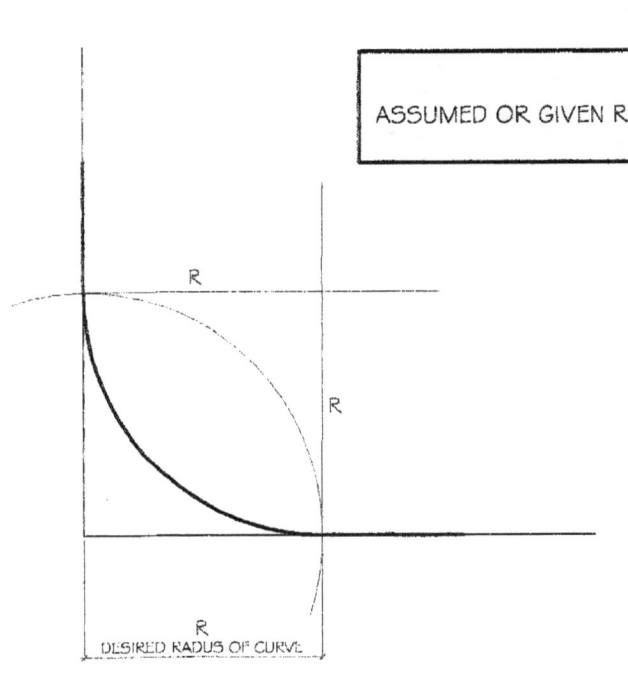

ARC TANGENT TO PERPENDICULAR LINES.

Offset: Inside + Inside

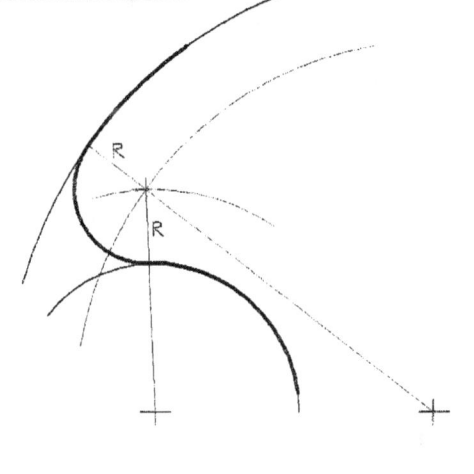

ARC TANGENT TO ARCS.

Offset: Outside + Inside

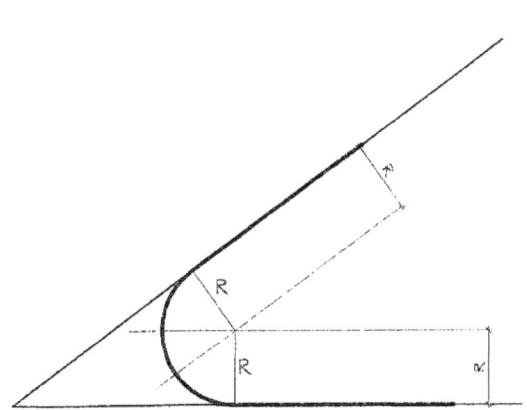

ARC TANGENT TO TWO LINES.

Offset: Inside + Inside

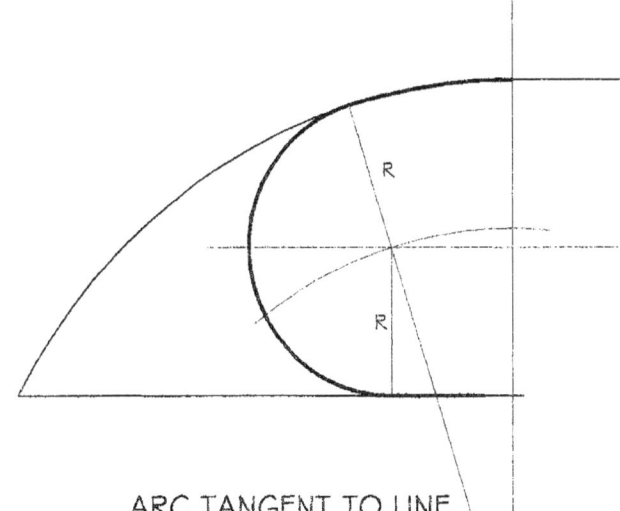

ARC TANGENT TO LINE AND ARC.

Offset: Inside + Inside

	ARC CONSTRUCTIONS	SCALE NTS	DWG
Mohawk College HAMILTON ONTARIO	NAME: Dorota M. Goede	DATE	T1

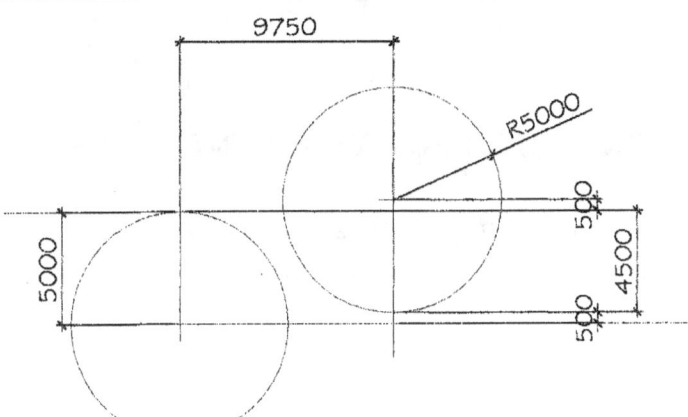

STEP 1:
DRAW CONSTRUCTION LINES AND CIRCLES PER DIMENSIONS SHOWN ON DWG C1.
GIVEN ARE TWO PARALLEL LINES AND TWO EQUAL CIRCLES.

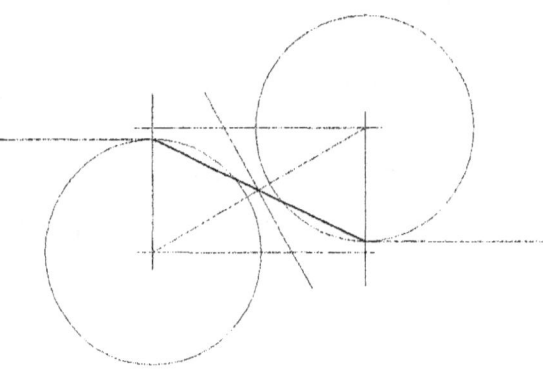

STEP 2:
CONNECT CIRCLE CENTERS. CONSTRUCT MIDLINES AS SHOWN.

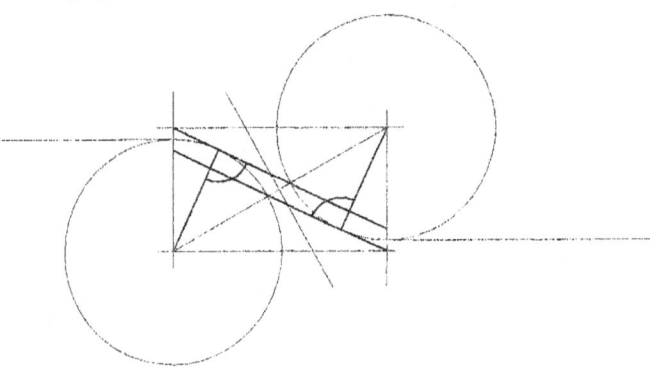

STEP 3:
DRAW PERPENDICULAR LINES AS SHOWN AND EXTEND TO ESTABLISH TANGENT POINTS.

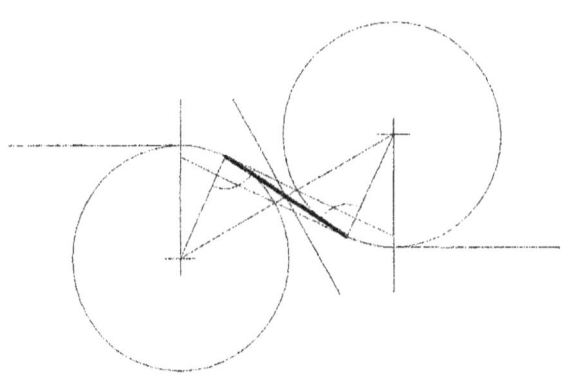

STEP 4:
DRAW FINAL, DARK, LINE FOR THE STRAIGHT PART OF THE ROAD BETWEEN TANGENT POINTS.

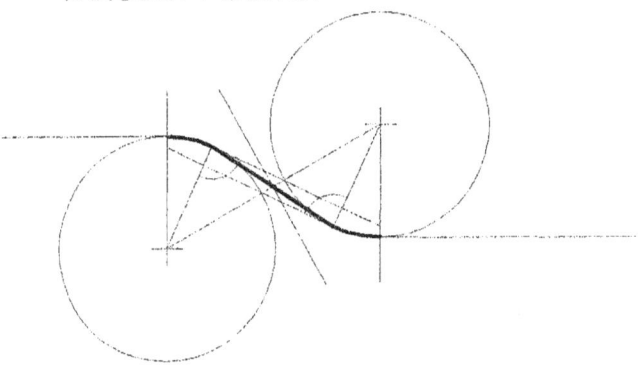

STEP 5:
CONSTRUCT CURVES, USING COMPASS, BETWEEN PRECISELY ESTABLISHED POINTS FOR THE BEGINNING AND END OF EACH CURVE.

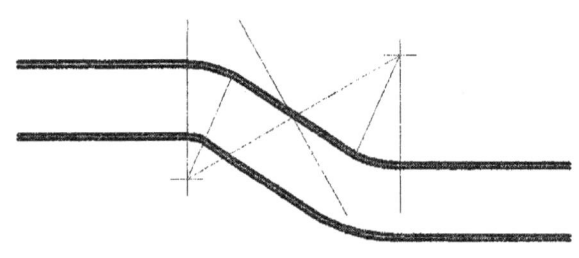

STEP 6:
COMPLETE CONSTRUCTION BY DRAFTING DARK THE FINAL, STRAIGHT STRECHES OF THE GIVEN ROAD. TO COMPLETE OFFSET BY 3000MM AND 200MM CURB

Mohawk College HAMILTON ONTARIO	ARC CONSTRUCTIONS	SCALE NTS	DWG
	NAME:	DATE	T2

ARC CONSTRUCTION ASSIGNMENT

Draw arc constructions shown on T1, T2 and T3 sheets. Use sheets in the handbook for practice and then draft precise constructions on A3 sheet in scale 1:100.

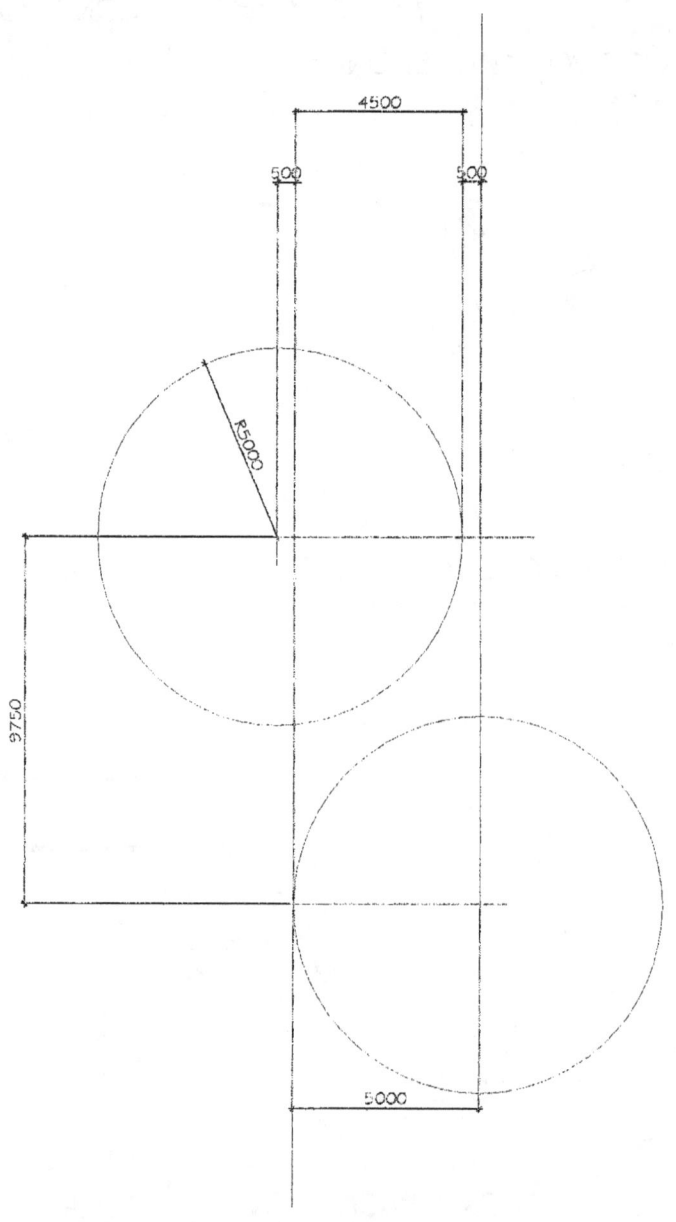

CONSTRUCTING AN OGEE CURVE BETWEEN TWO PARALLEL LINES.

GIVEN

STEP 1

STEP 2

STEP 3

STEP 4

STEP 5

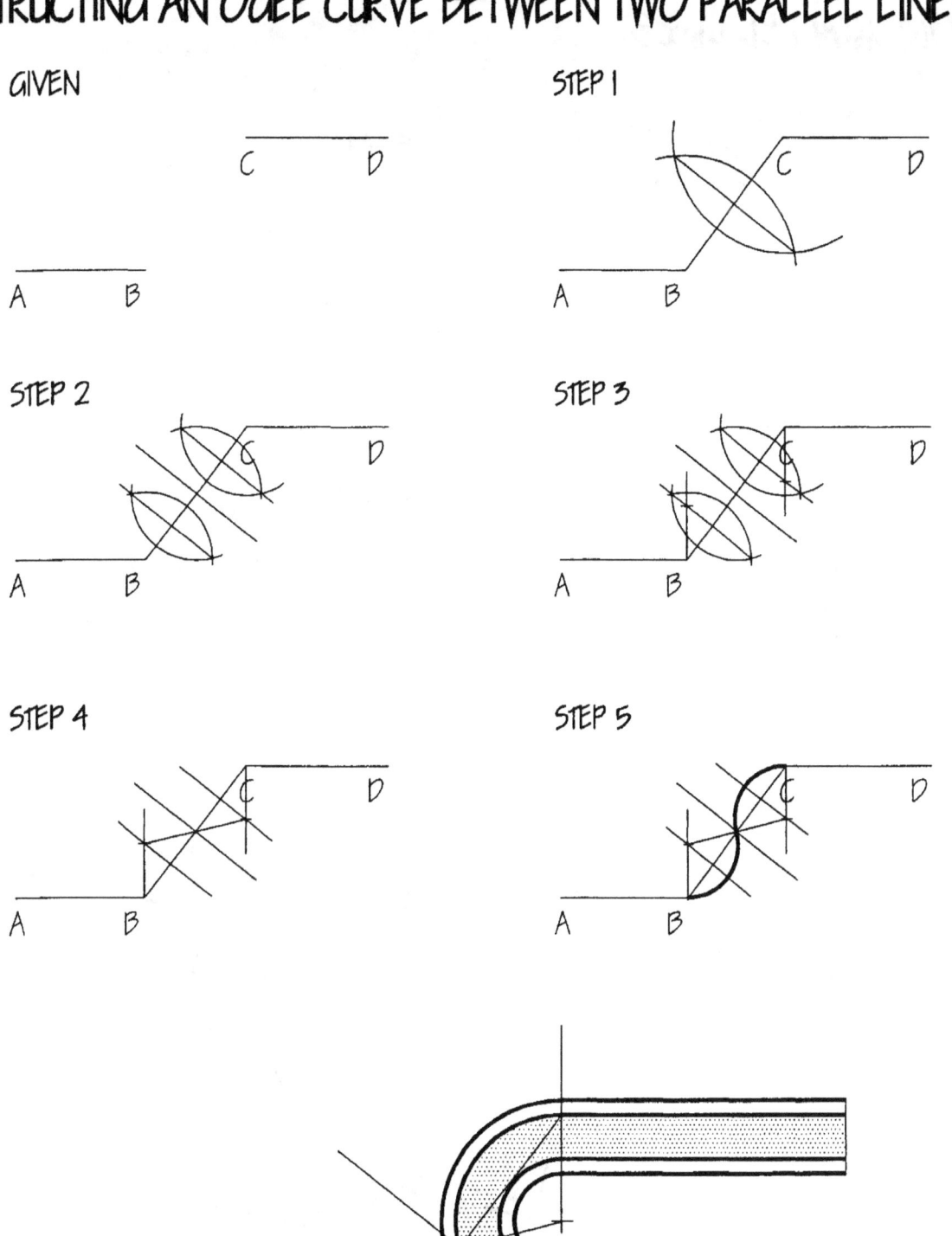

APPLICATION OF OGEE CURVES IN ROADWAY DESIGN.

CONSTRUCTION OF AN OGEE CURVE BETWEEN TWO NONPARALLEL LINES.

GIVEN

STEP 1

STEP 2

STEP 3

STEP 4

STEP 5

APPLICATION OF OGEE CURVES IN ROAD DESIGN.

July 95

LOCATING THE CENTER OF A CIRCLE USING CHORD METHOD:

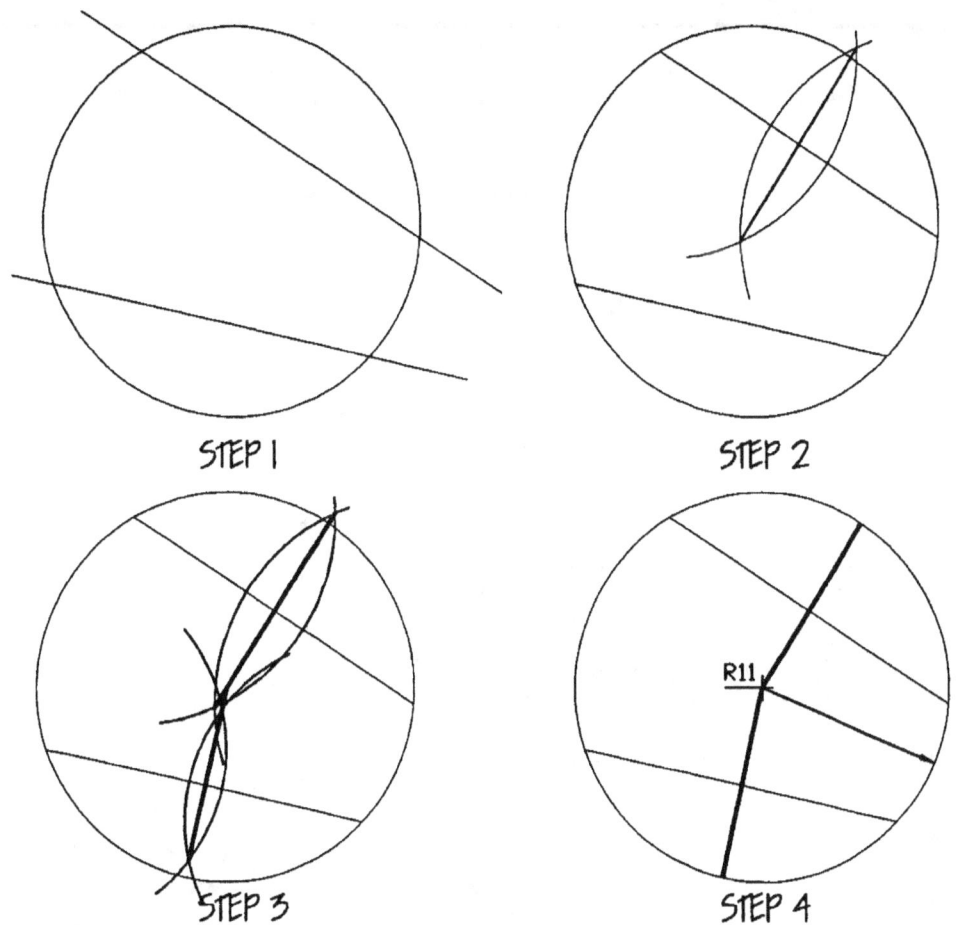

CONSTRUCTING IRREGULAR CURVES OF ARCS BY DRAWING A SERIES OF TANGENT ARCS:

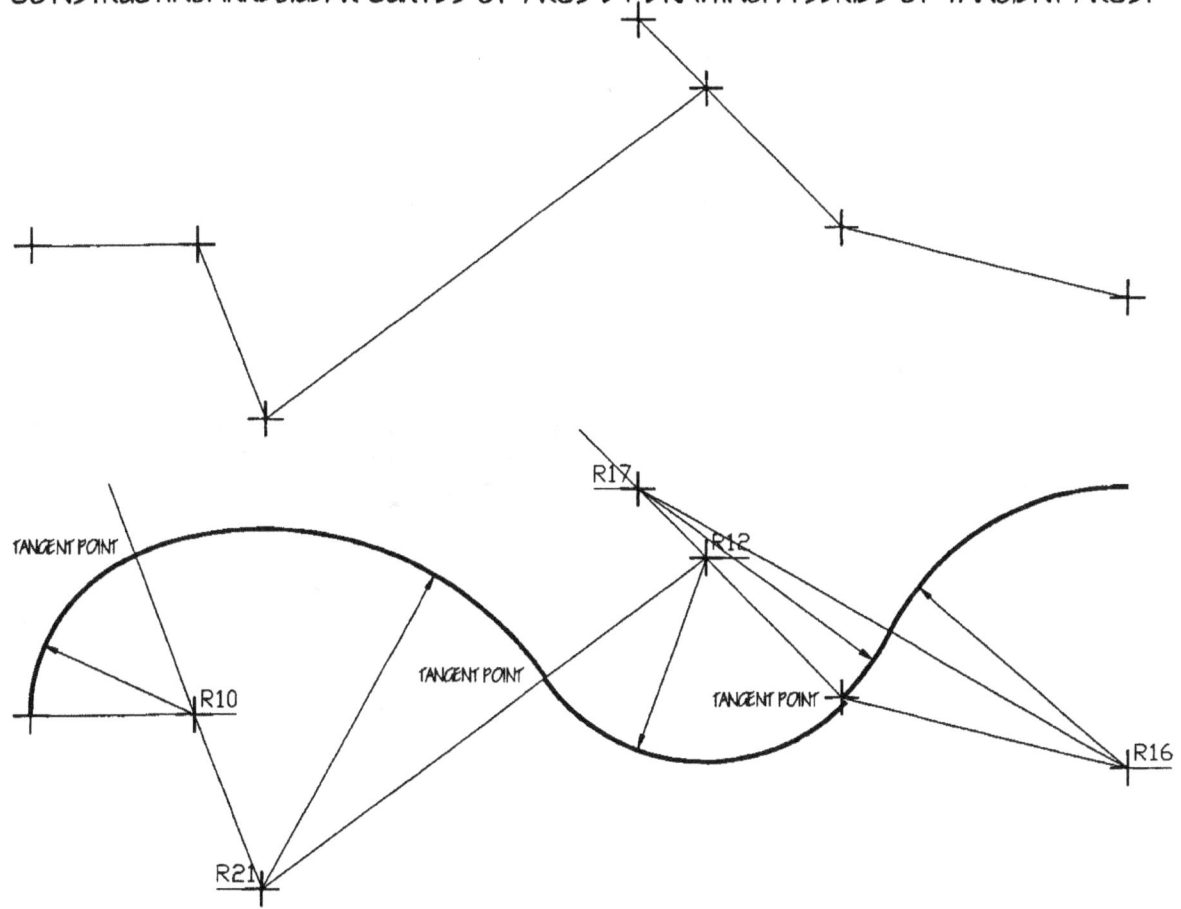

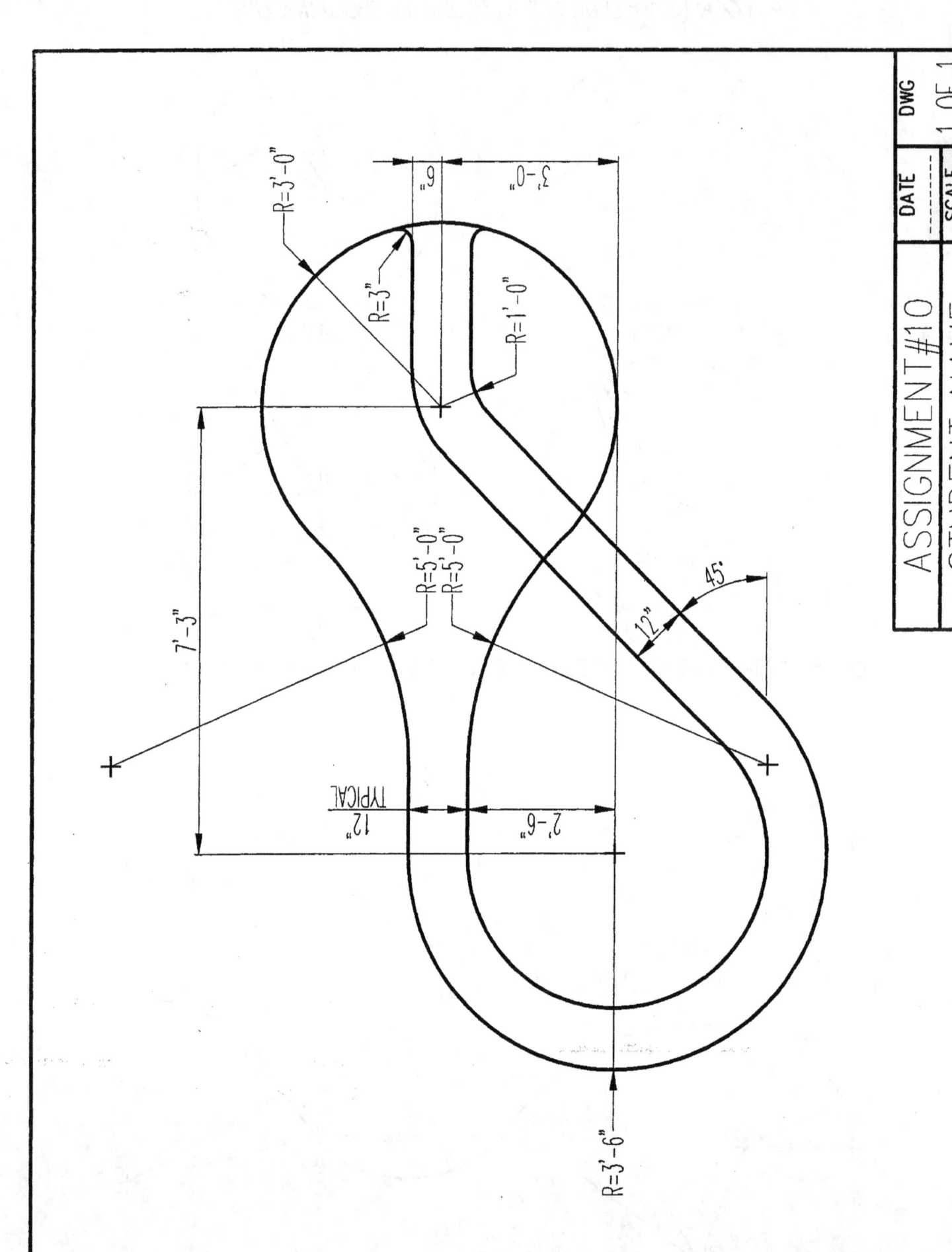

DAVERN INTRSECTION PROJECT

The Regional Municipality of Davren has sent their crack fieldsman, Mr. P. T. Barnum, out to the intersection of Hunt Road and Adams Street to draw some sketches of the intersection and the roadway cross-sections. The results of P. T. Barnum's visit to the site are shown on the attached page. P. T. sketched a plan of the intersection showing the lane arrangement, curb radii, stop bars, sidewalks, and fences. He also sketched two cross-sections, one of each roadway, showing the important features and dimensions of each road.

The drawing has landed on your desk, and your task is to transform the field notes into a scale drawing of the intersection, which will be used later for traffic control signal design. On an A2 sheet at a scale of 1:200, draw the plan view of the intersection using the information from the field notes. The plan should show (and dimension) all the important features of the roadways and intersections: lane width and arrangement; stop bars; property lines; center lines; lane lines; fences; sidewalks; and boulevards.

Use a Type B Title Block (vertical on right side and 50 mm wide) and draw the portions of roads within the following km stations:

    1+940 to 2+060 Hunt Road
    5+425 to 5+575 Adams Street

The road chainages at the intersection of centerlines are:

    2+000.000   Hunt Road = 5+500.000 Adams Street

Evaluation will be based on the accuracy and completeness of the drawing, the dimensioning and labeling, and of course, your drafting techniques.

Note: Remember, build out from the center line

FIELD SKETCHES: HUNT ROAD @ ADAMS STREET

PLAN VIEW:

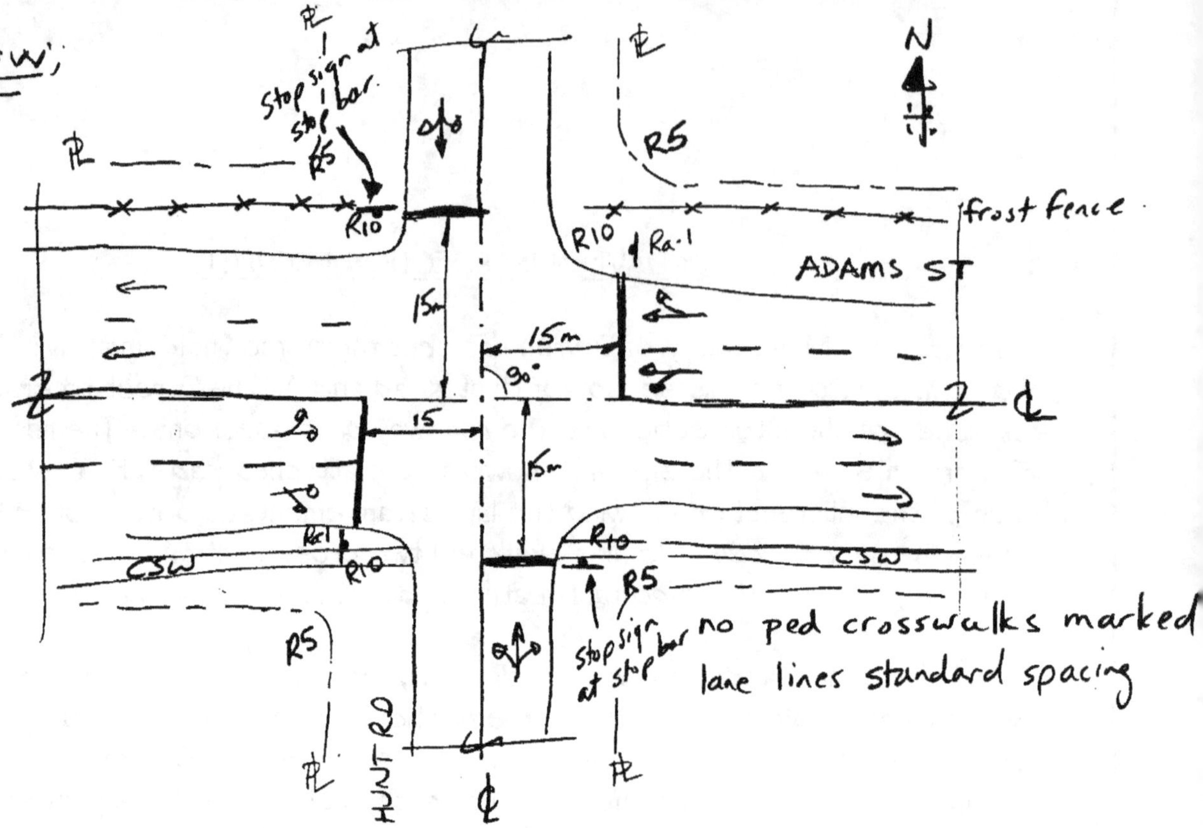

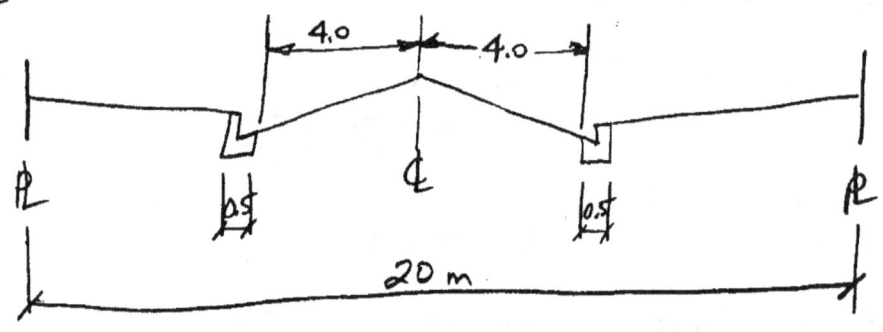

no ped crosswalks marked
lane lines standard spacing

CROSS-SECTION: HUNT RD

CROSS SECTION: ADAMS ST.

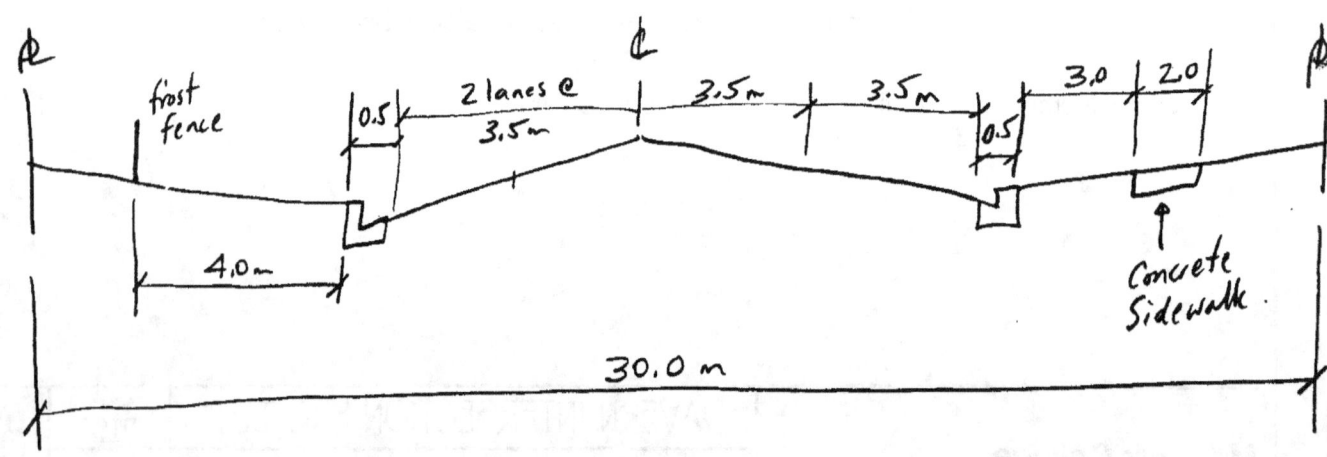

FIELD NOTES BY: P.T. BARNUM   91/12/02

SCHOOL SIGNAL AND CROSSING.

Use drawing T1 as a layout guide for this project. Work on an A2 size sheet and draw the intersection in 1:200 scale.
You will need a quality compass in order to complete this assignment successfully. Optionally, you may use a small symbol template containing squares, circles etc., and the drafting lead sharpening paper pad. The pad consists of sheets similar to a very fine sanding paper. It is used for compass lead sharpening.

Start with the most important construction lines and outline your road geometry. Use the information printed in larger scale on construction sheets C3 and C4. The construction lines are also shown to assist you.
Note that dimension lines are continuous and separate dimensions line up along a construction line.

All curves are drafted using precise constructions for the start and end points of each curve. Never draw dark object straight lines first and then join them with a curve. Draw construction lines first, establish the beginning and end points of a curve, and then draw the curve using a compass. The dark, object lines, connecting the ends of curves are drawn last. Match the pencil lead and the line thickness of your compass and pencil. Draw lines precisely from the ends of the curve.
Your Instructor may assign the C1 and C2 drawings to be completed separately as an introductory geometry drafting assignment. The construction worksheets are included.

The road lines are all cut off precisely at a light construction line. Establish the location of these construction lines to suit the scale and layout of your drawing. Allow drafting space for the dimension lines to be added later.

The traffic signal head and pole symbols are best drafted using a technical template containing small circles and triangles. The elaborate computer bus symbol should be simplified and shown as a rectangular box (12m X 3m) or simplified and traced from the C3 sheet.

Frame the completed drawings with a standard Mohawk College titleblock.

| Mohawk College | SCHOOL CROSSING | SCALE NTS | DWG |
| HAMILTON ONTARIO | NAME: Dorota M. Goede | DATE Jan.02ᵀ | INSTRUCTIONS |

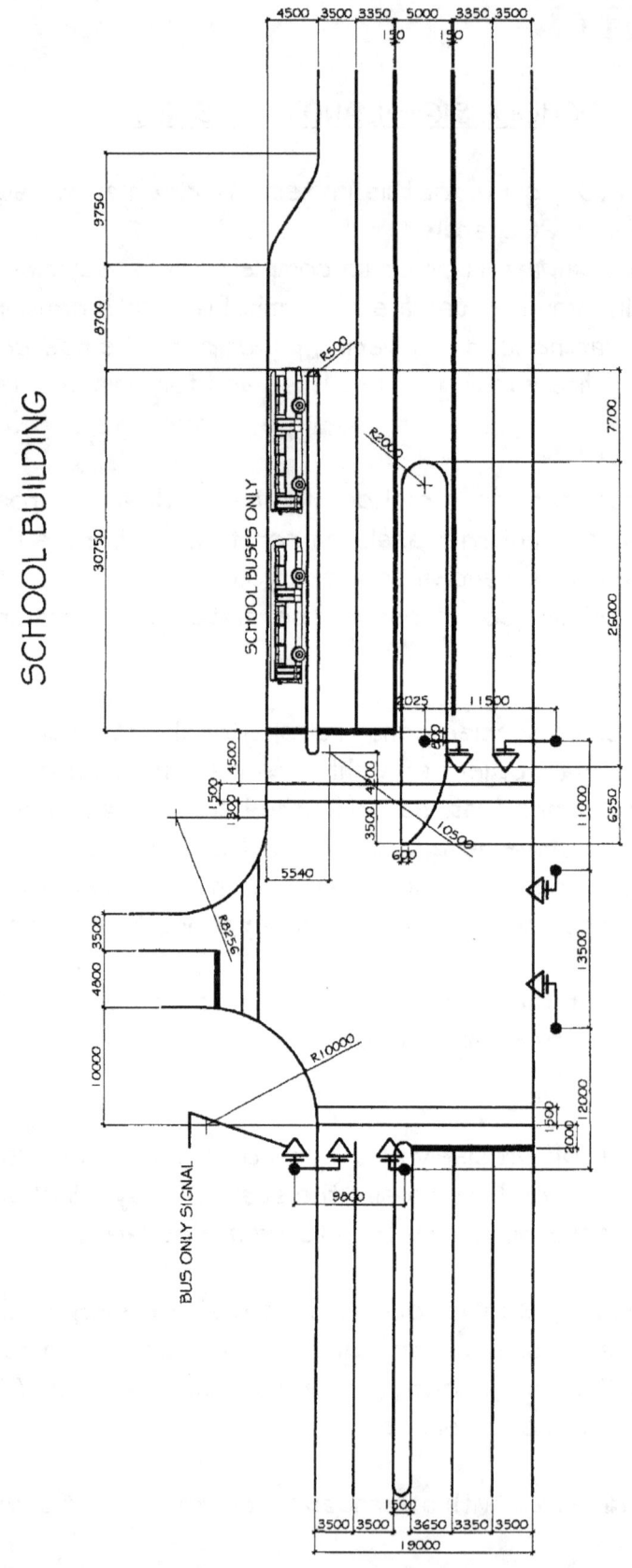

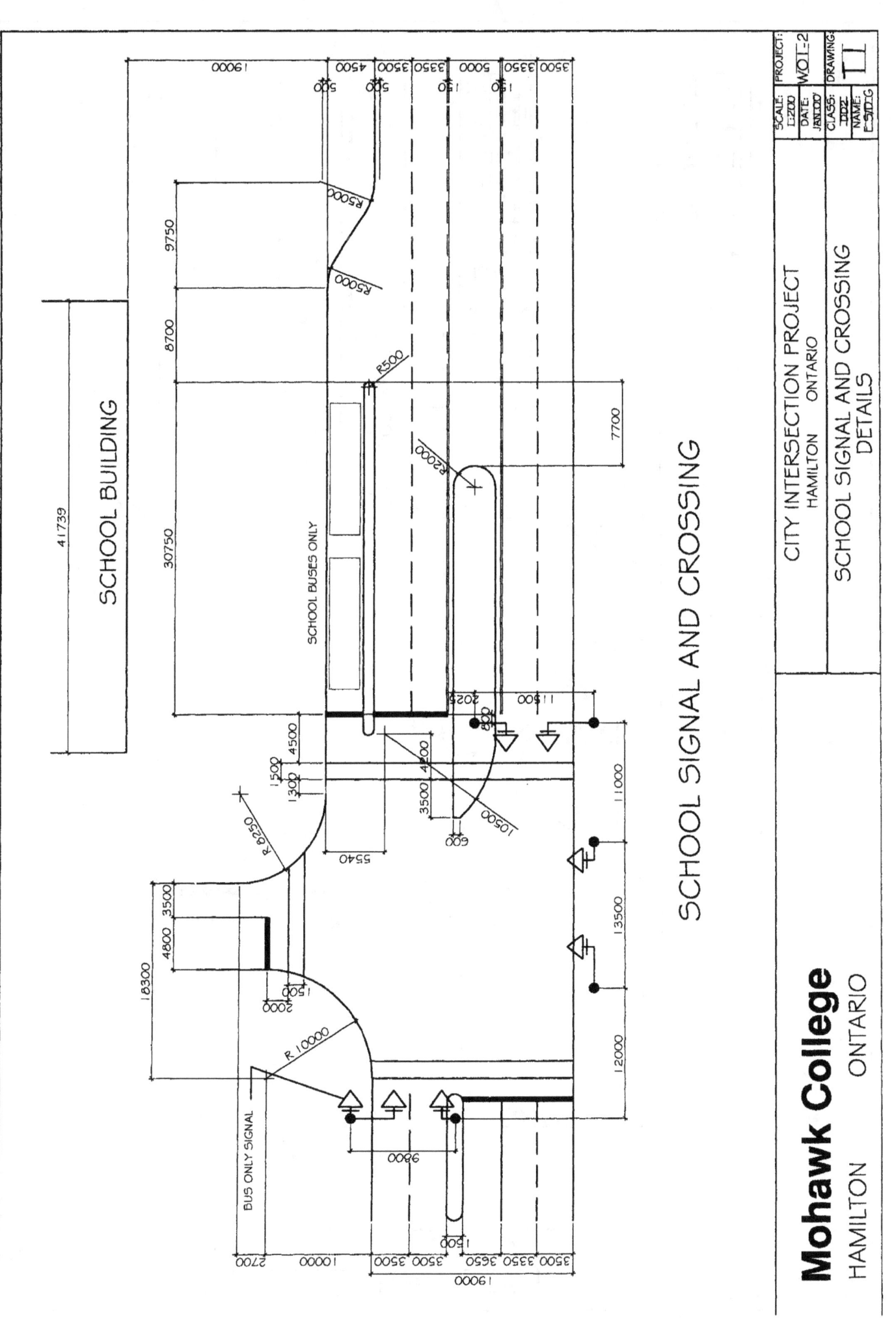

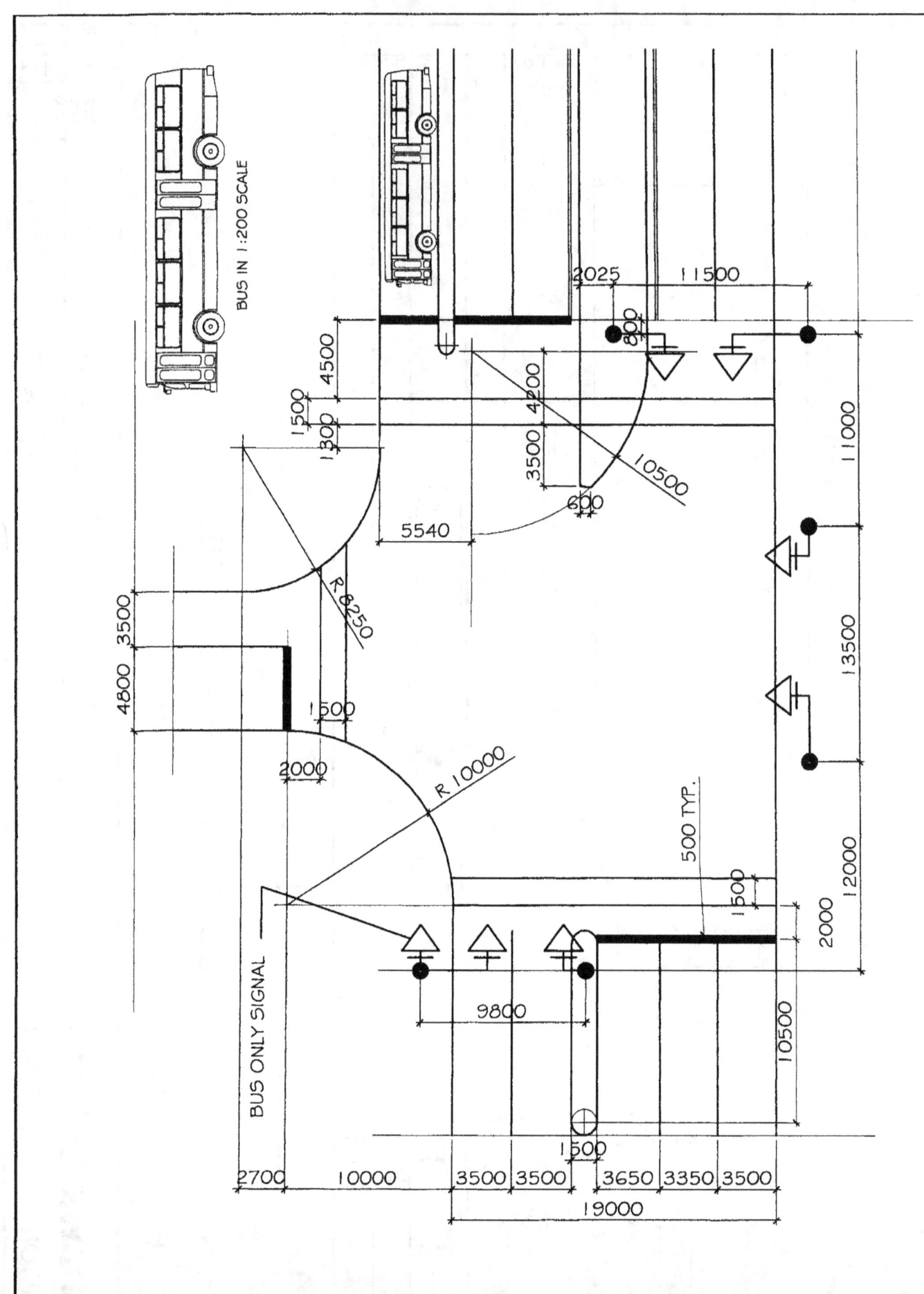

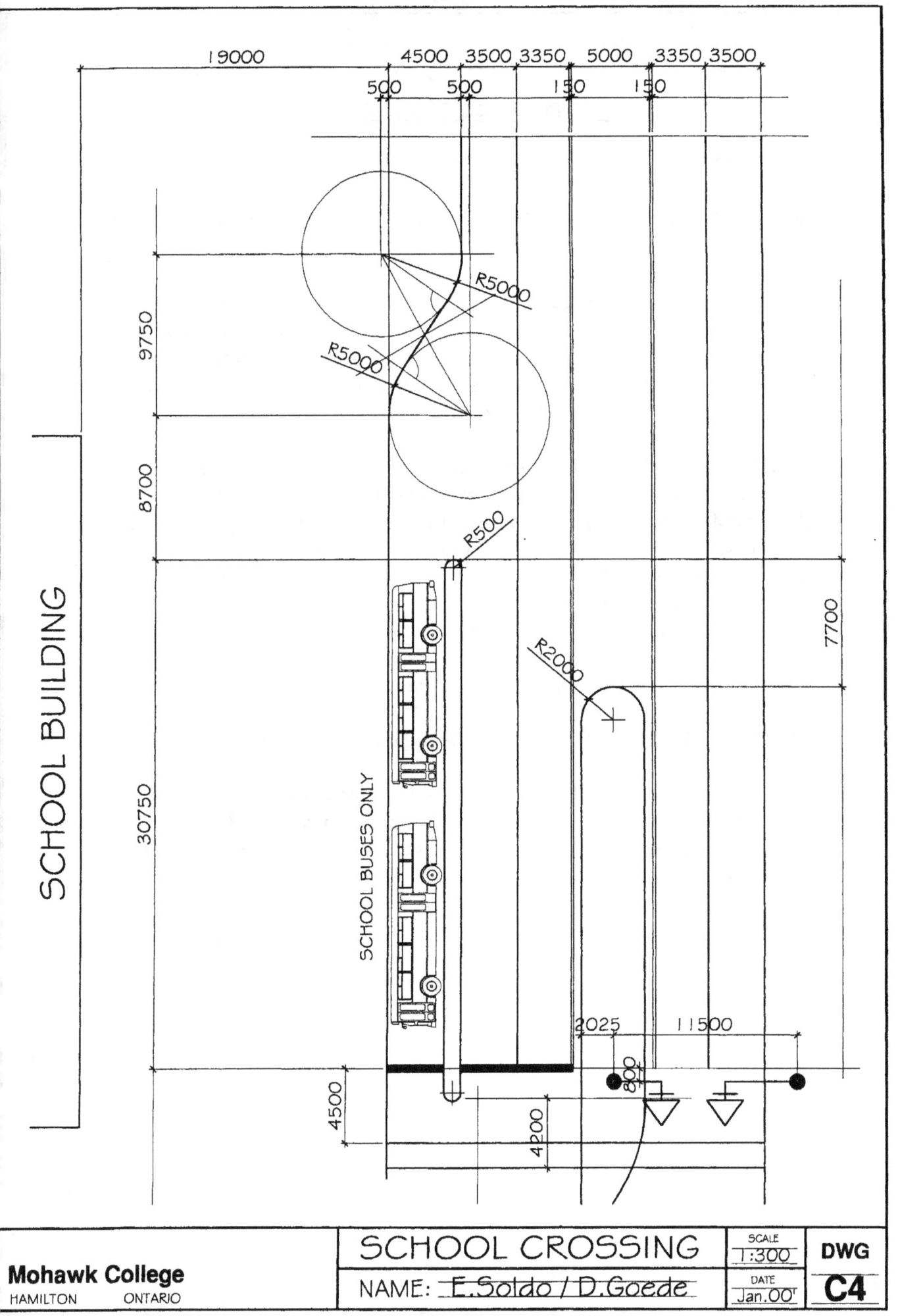

SITE PLAN DRAFTING AND DESIGN

ASSIGNMENT 1:

The handout sheet L1 represents typical Site Plan information for the preliminary Site Plan proposal drawings.

Draw all existing site information on A2 size sheet. Use 1:500 metric scale. Layout your sheet by outlining the main street and all property lines. Draw vertical and horizontal property lines first. For angled lines use your protractor. You may also use the handout to construct the property lines that are angled. Use both triangles and draft parallel lines in desired locations.

Pay special attention to quality and precision of your lines. Start with light construction lines and locate all crucial information. Show property lines and building roof outline using darkest lines; use a medium line weight for notes, trees, roads and other site details. Draw site contour lines freehand and very light. Use dimensions shown to locate the contour lines as close as possible to the handout information. You may draw every other contour line if lines seem congested. Remember to use minimum 3 line weights.

Design your own titleblock per Mohawk College standards.

ASSIGNMENT 2: FINAL PROJECT FOR TRANSPORTATION STUDENTS

Design road access to and from the property. Layout and maximize the parking area. Provide access and maneuvering space in designated service area.
Refer to Architectural Graphic Standards textbook for all necessary design information.
Design and show your proposal for landscape design.
Render your proposal to emphasize green spaces, building lot percentage coverage and asphalt areas.

Mohawk College
HAMILTON   ONTARIO

DUNDAS PROFESSIONAL BUILDING	SCALE NTS	**DWG**
NAME: Dorota M. Goede	DATE Dec. 01'	INSTRUCTIONS

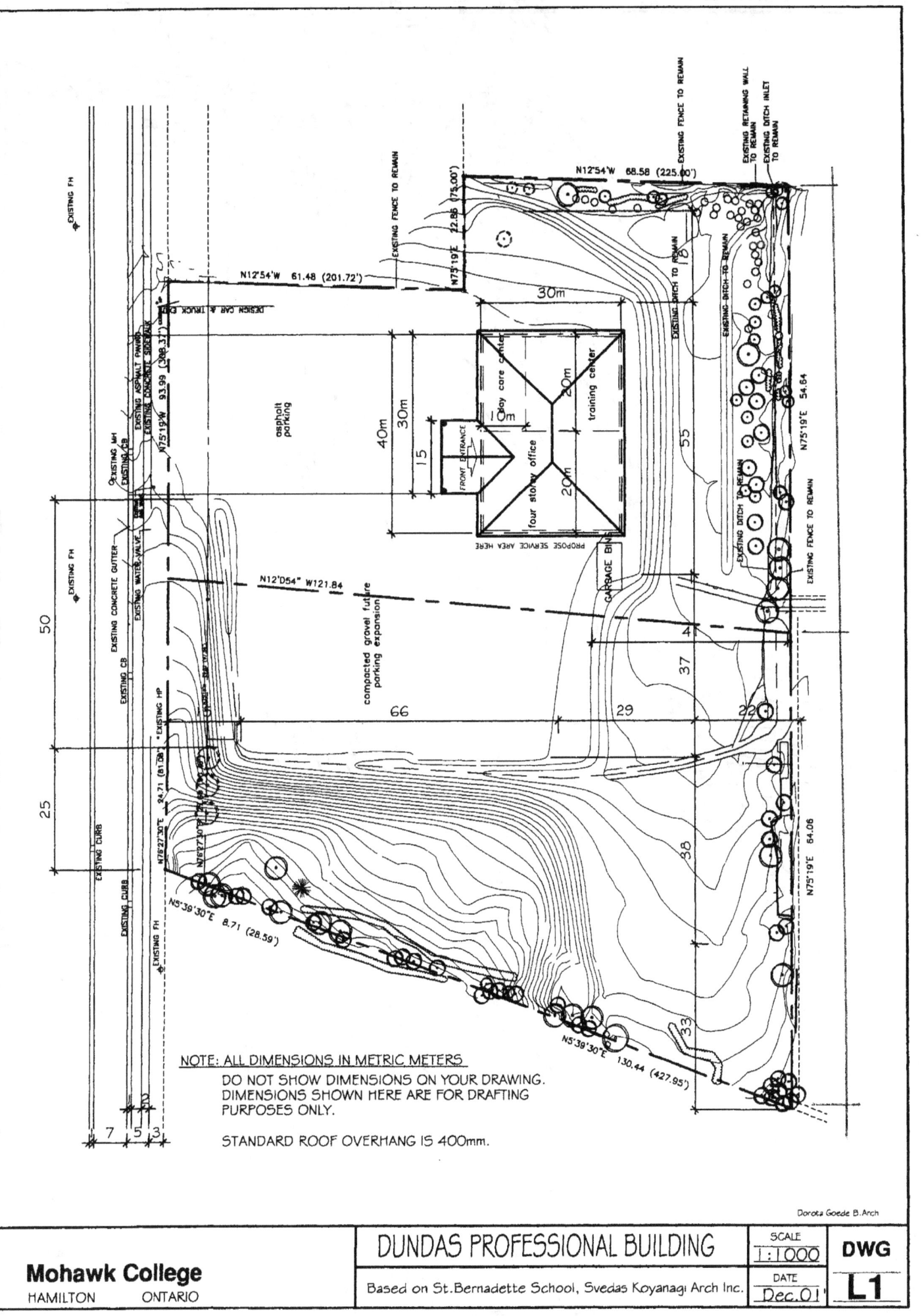

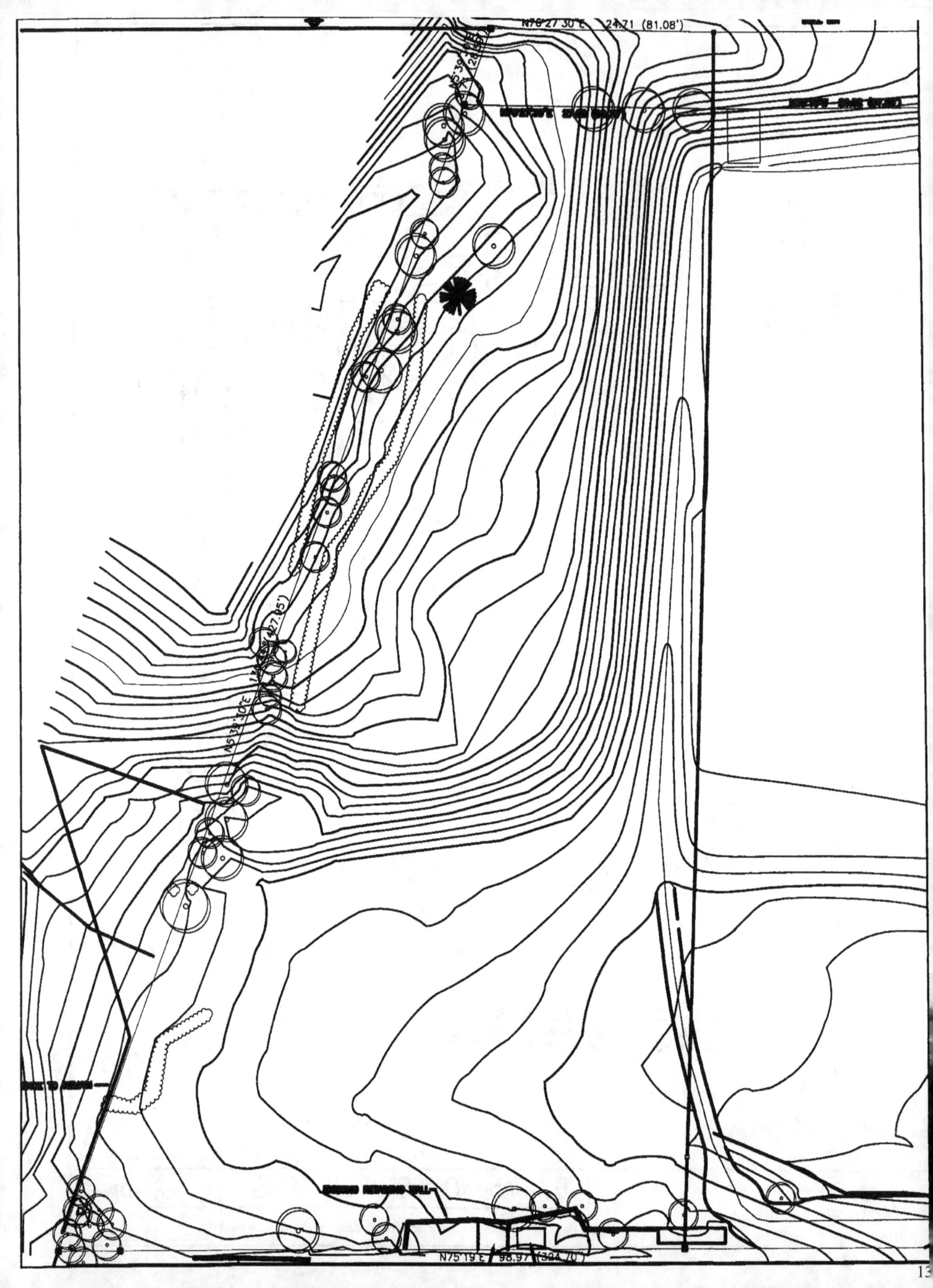

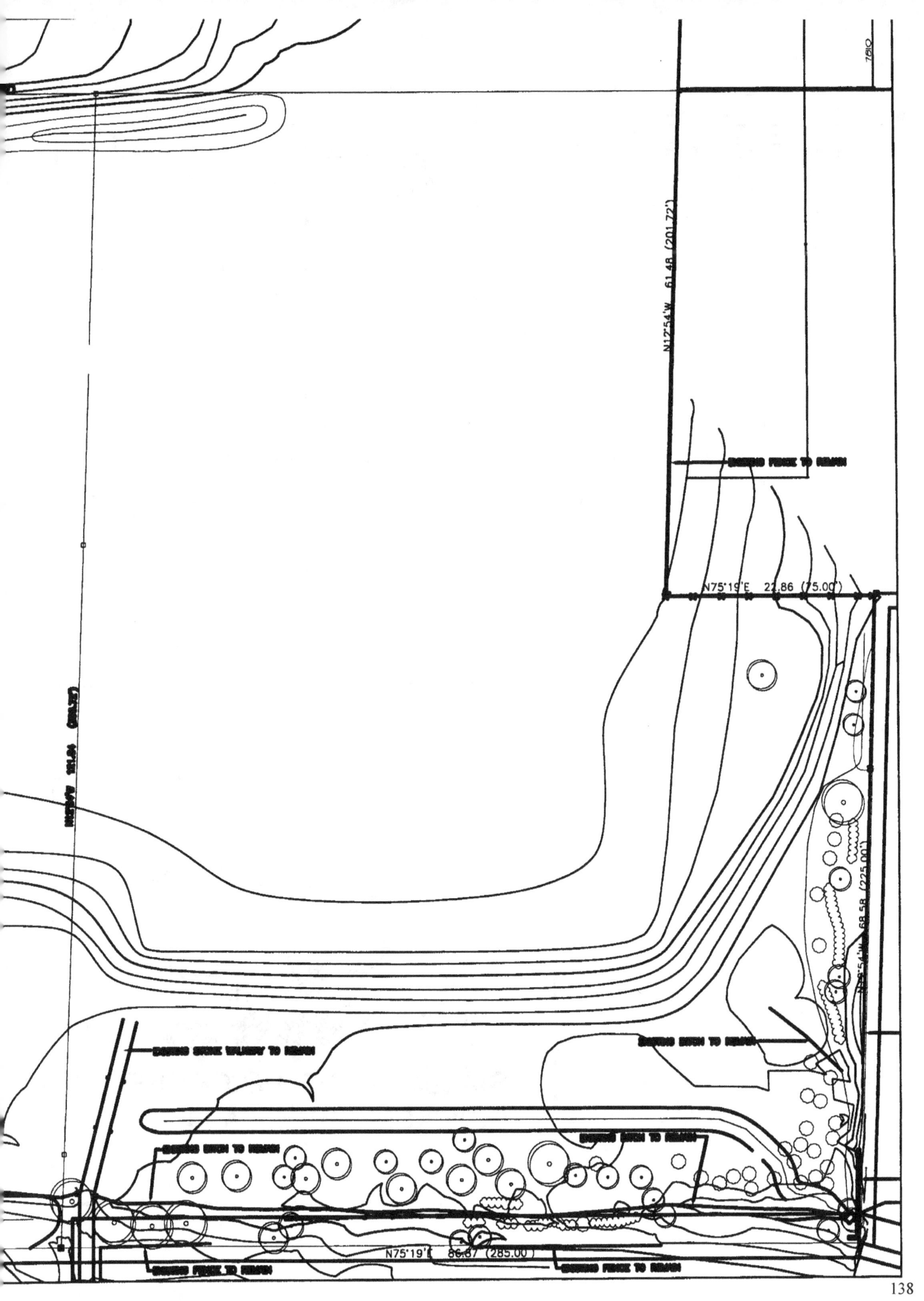

LEGEND

○FH	FIRE HYDRANT
○	TREE
□CB	CATCH BASIN
➡	TRAFFIC ARROW
♿	HANDICAP SPACE
▦	CONCRETE PAVERS
▨	CONCETE WALK
〰️	TREE LINE
☼LP	LIGHTS
☼FP	FLAG POLES
⊙LP	MOONRAYS (LIGHTS)
●BL	BOLLARDS

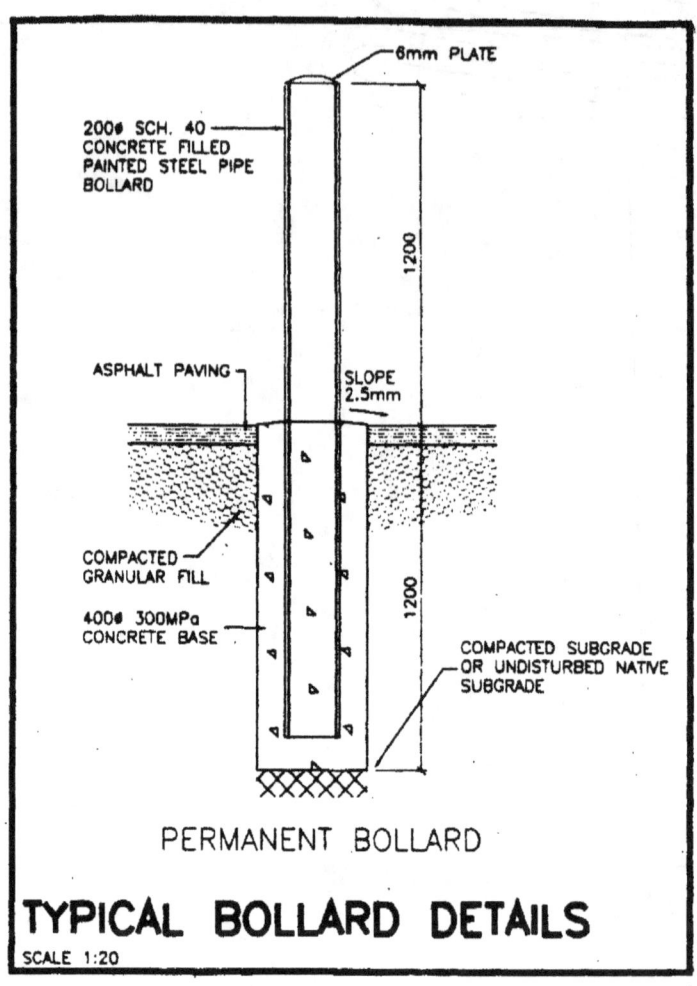

PERMANENT BOLLARD

TYPICAL BOLLARD DETAILS
SCALE 1:20

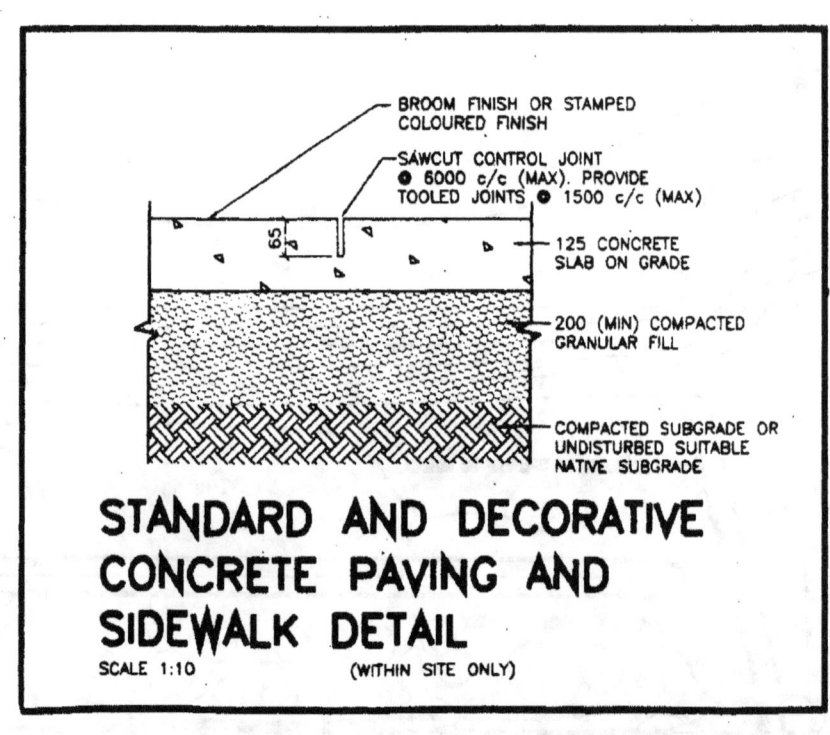

STANDARD AND DECORATIVE CONCRETE PAVING AND SIDEWALK DETAIL
SCALE 1:10   (WITHIN SITE ONLY)

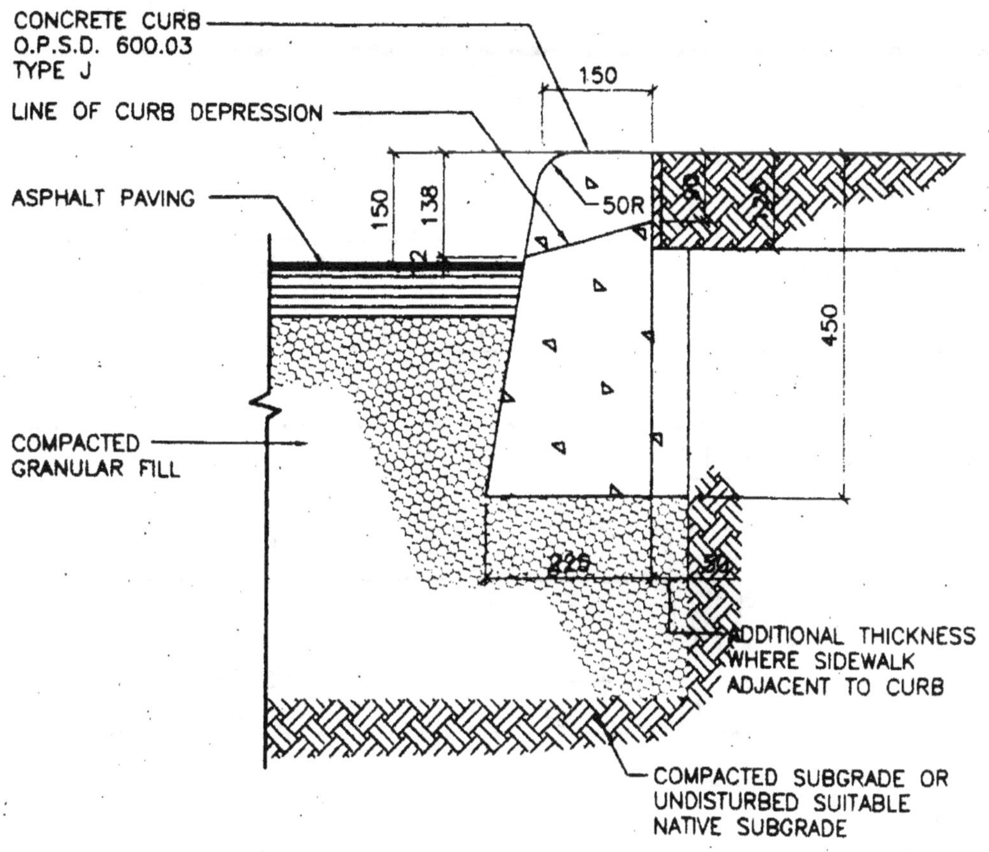

CONCRETE CURB DETAIL
SCALE 1:10 (WITHIN SITE ONLY)

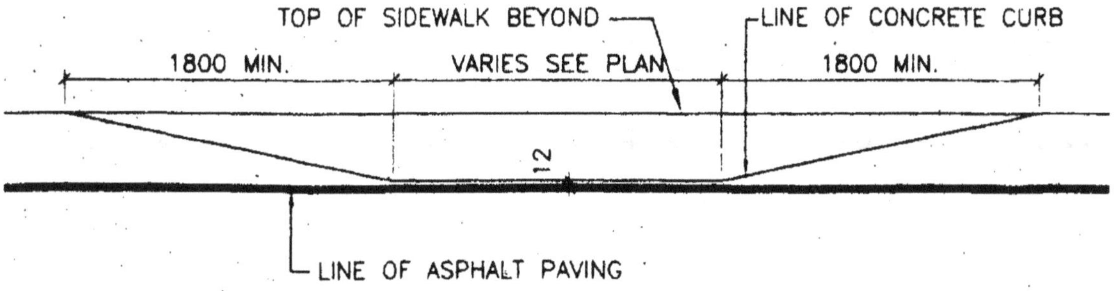

TYPICAL CURB DEPRESSION
NOT TO SCALE

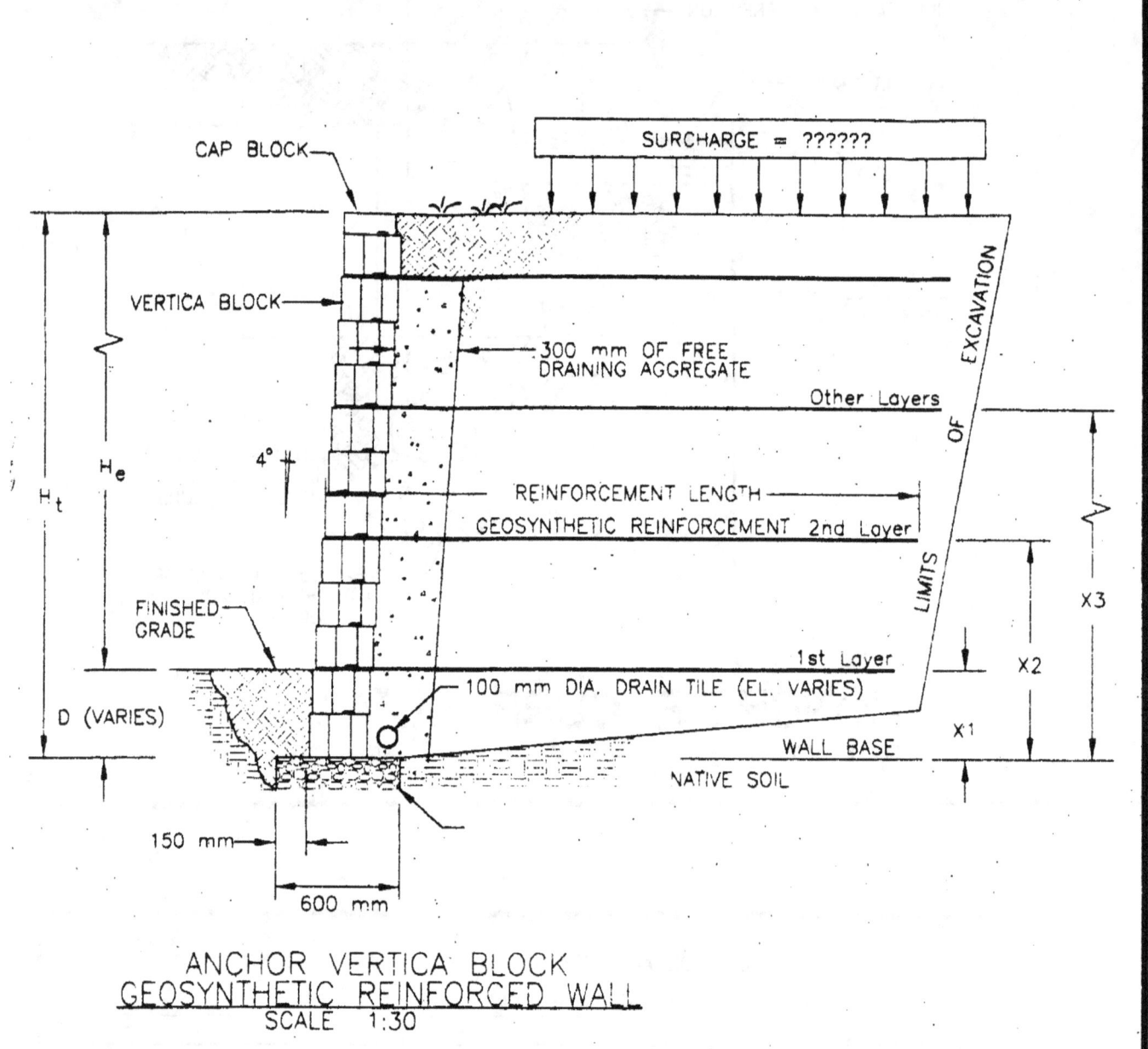

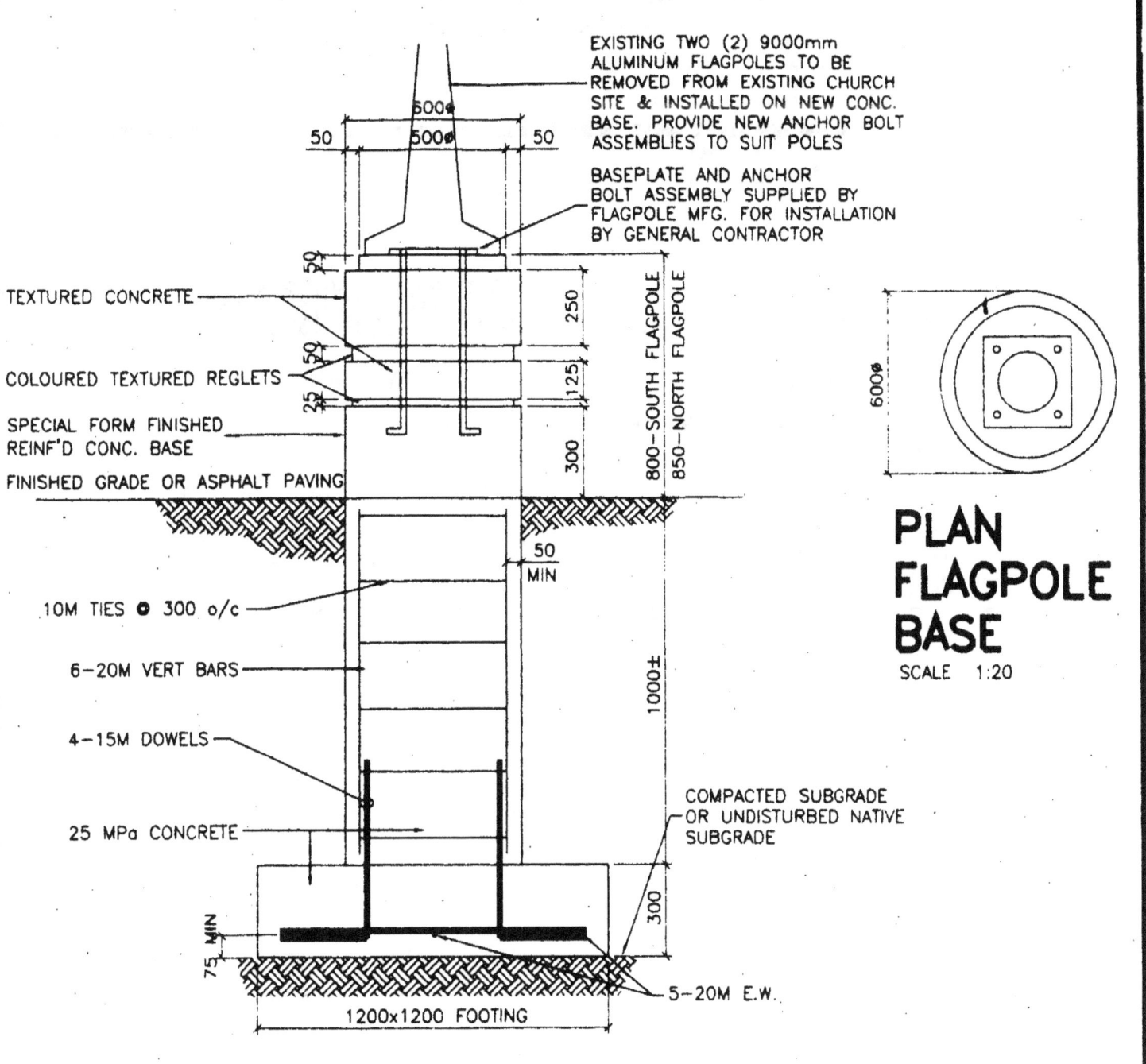

Department of
BUILDING AND CONSTRUCTION SCIENCES
Mohawk College

GENERIC SECTION APPENDIX & SIMBOLS

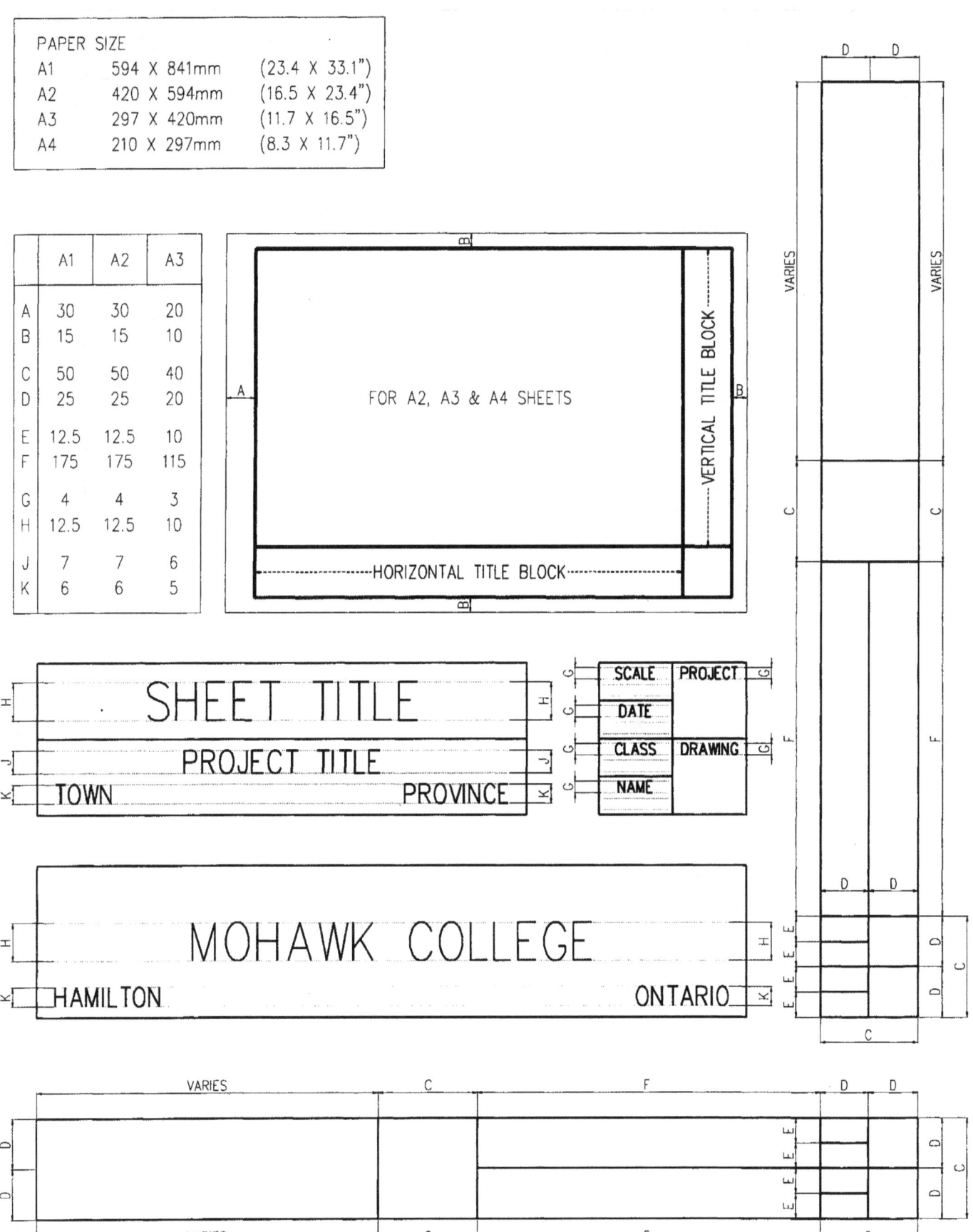

PROJECT	DRAWING
SCALE	DATE	CLASS	NAME

SHEET TITLE

PROJECT TITLE

TOWN   PROVINCE

MOHAWK COLLEGE

HAMILTON   ONTARIO

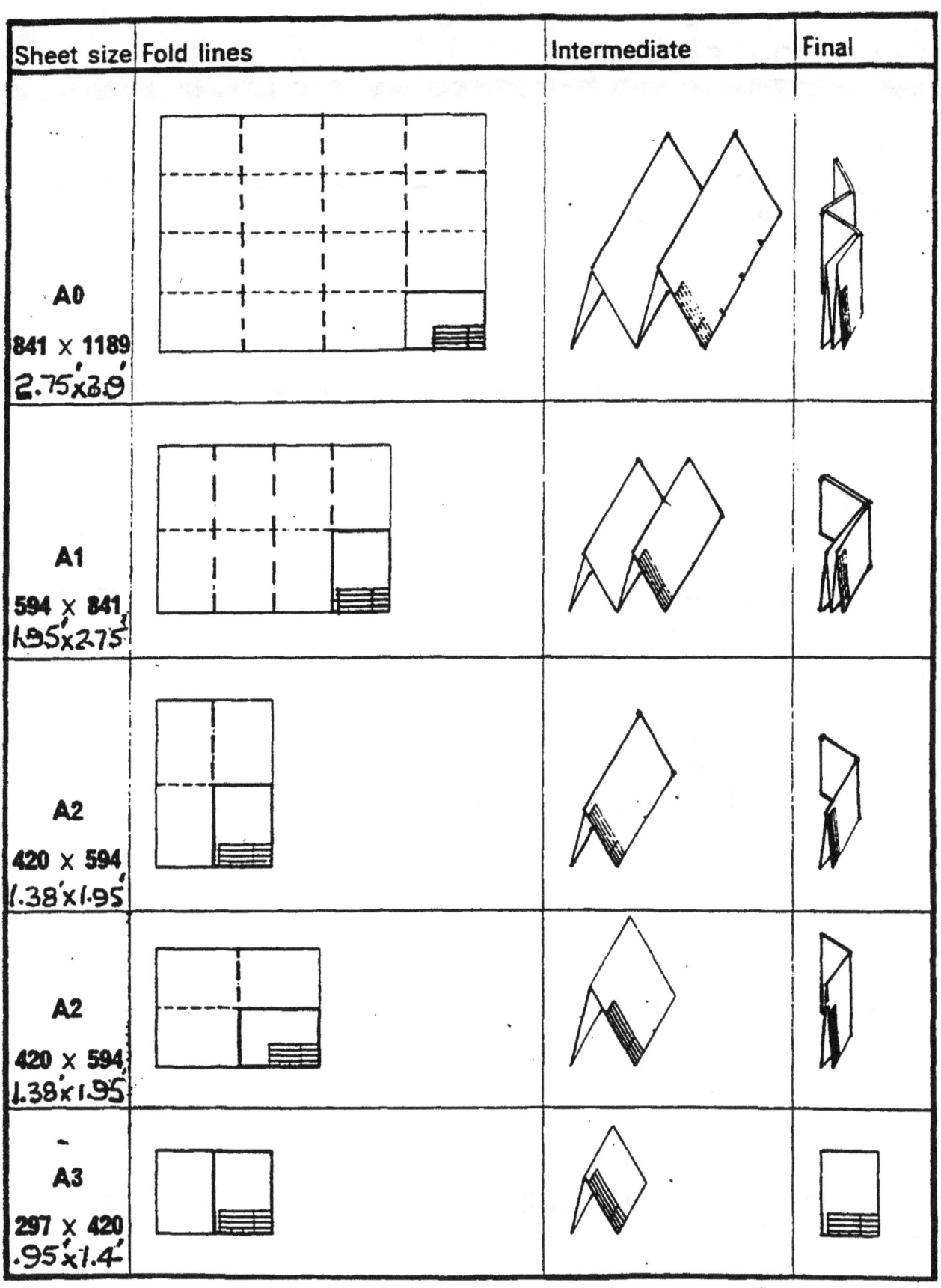

Manual on Metric Building Drawing Practice (1977)

LINE CONVENTIONS

VISIBLE OBJECT LINES ———————————————— (MEDIUM)
PEN POINT EQUIVALENT IS 0·6 (U.S.A. 2·5)

HIDDEN LINES ------------------- (MEDIUM)
P. P.E. IS 0·6

CENTER LINES ——— · ——— · ——— · ——— (THIN)
P. P.E. IS 0·2 (U.S.A. 00)

REFERENCE OR DATUM LINES ——— — ——— — ——— (THIN)
P.P.E. IS 0·2

SECTION LINES ═══════════════════ (THICK)
P.P.E. IS 1·2 (U.S.A. 4)

DIMENSION & EXTENSION LINES ———————————— (THIN)
PEN POINT EQUIVALENT IS 0·2

Revised June 1978

4

SYSTEMS BUILDING

BASIC MODULAR DRAFTING CONVENTIONS

BASIC MODULE : M = 100 mm

Dimension from the main (structural and/or planning) module (expressed always in modules) (and located within hexagon box).

Other dimensions: 45°-degree arrows for dimensions from any modular range.

Dots for spacing and/or dimensions between components located off grid, otherwise arrow.

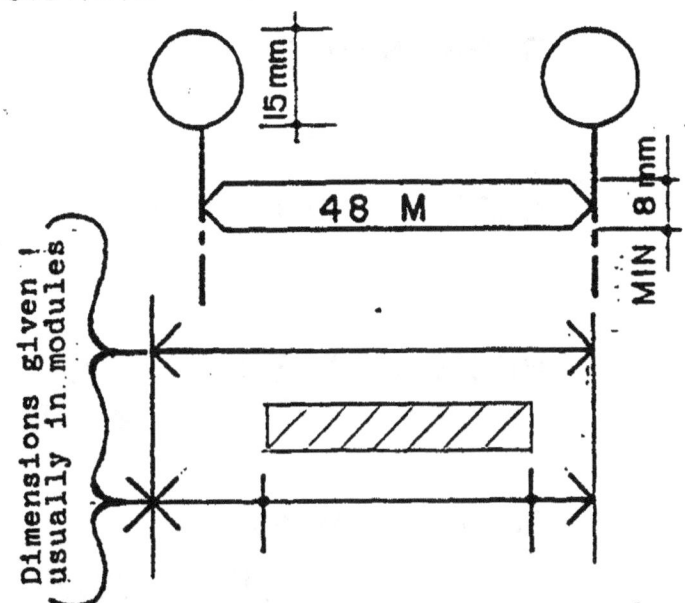

MODULE LINE CONVENTIONS :

Statical-structural module

Planning module

Basic module (M)

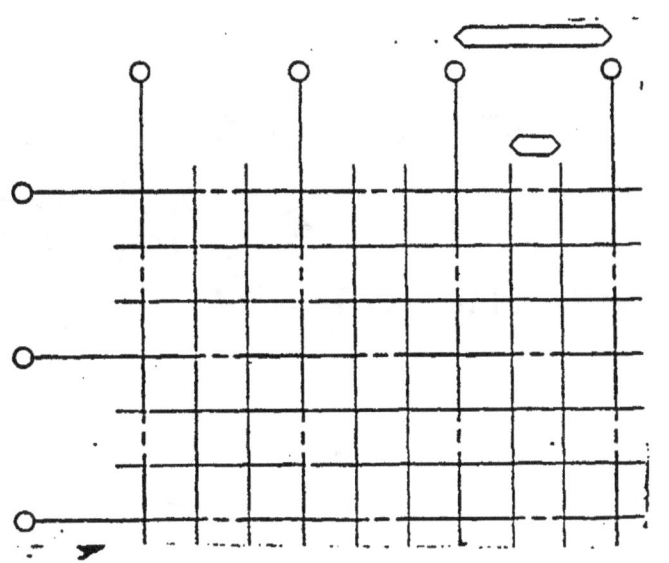

DRAWING GRID

Revised June 1978

6 114

DRAFTING SYMBOLS MATERIALS & ABBREVIATIONS

Follow the following symbols in addition to the drafting symbols, materials and abbreviations as prescribed in the "Manual on Metric Building Drawing Practice", pages 39-63 inclusive.

TITLES — ELEVATION

ROOM NAMES — CLASSROOM

NOTES & DIMENSIONS — ALUMINUM SCREEN

ROOM NUMBERS — 103

DOOR NUMBERS — 105

DETAIL REFERENCE — 12 / DETAIL NO. SEQUENCED / .6 / 3 DWG. NO. — 5/25

COLUMN REFERENCE — FOR REF'CE — B

*** ARROWHEAD INDICATIONS** — DIMENSION LINES 10, 20 — EXTENSION LINES — 20 | 345

SECTION SYMBOLS — SECTION NO. / DRAWING NO. — A/3 BLDG. SECTION — 2/5 WALL SECTION

NORTH POINT — NORTH

REVISED JUNE 1978   note: dimensions in millimetres

* CONSULT WITH APPROPRIATE INSTRUCTOR

MOHAWK COLLEGE
BUILDING & CONSTRUCTION SCIENCE

PROFILE LINE	HEAVY LINE WEIGHT
OBJECT LINE	MEDIUM LINE WEIGHT
HIDDEN LINE	INDICATES OBJECT NOT WITHIN IMMEDIATE VIEW IE ABOVE OR BELOW CUTTING PLANE
BREAK LINE	REPRESENTS TERMINATION OF OBJECTS EXTENDING BEYOND THE IMMEDIATE VIEW
GRID LINE	LINE OF REFERENCE FOR LOCATION OF ELEMENTS. USUALLY CENTRE OF COLUMN OR FACE OF A WALL
CENTRE LINE	INDICATES THE CENTRE OF AN OBJECT
DIMENSION LINE	USED TO INDICATE SIZE OR LOCATION OF AN OBJECT
LEADER LINE	CONNECTS A NOTE OR A DIMENSION TO A POINT ON THE DRAWING
GUIDE LINE	USED FOR LAYING OUT OR CONSTRUCTING A DRAWING. THEY ARE NOT INTENDED TO BE SEEN.
PROPERTY LINE	INDICATES THE EXTENT OF A PROPERTY.
UTILITY LINE	INDICATES THE LOCATION OF AN UNDERGROUND SERVICE.

DRAW INFORMATION SHOWN, SCALE 3/8"=1'-0", ADD MISSING DIMENSIONS, IDENTIFY EACH TYPE OF LINE WITH A LEADER LINE AND NAME

CLASSIFICATION OF LINES

THICK BLACK LINES
1. BORDER LINES
2. TITLE UNDERLINE
3. CUTTING PLANE LINE

MEDIUM BLACK LINES
4. OBJECT LINES
5. HIDDEN LINES

THIN BLACK LINES
6. DIMENSION LINES
7. EXTENSION LINES
8. CENTRE LINES
9. LEADER LINES
10. BREAK LINES
11. SECTION LINES (HATCH)

FAINT (LIGHT) GREY LINES (ERASE LONG ENDS)
12. GUIDE LINES FOR LETTERING
13. CONSTRUCTION LINES.

RETAINING WALL

SECTION A-A
SCALE 1/4"=1'-0"

BUILDING DRAWING PRACTICE
DIMENSION LINES
(THIN BLACK LINES)

**① **

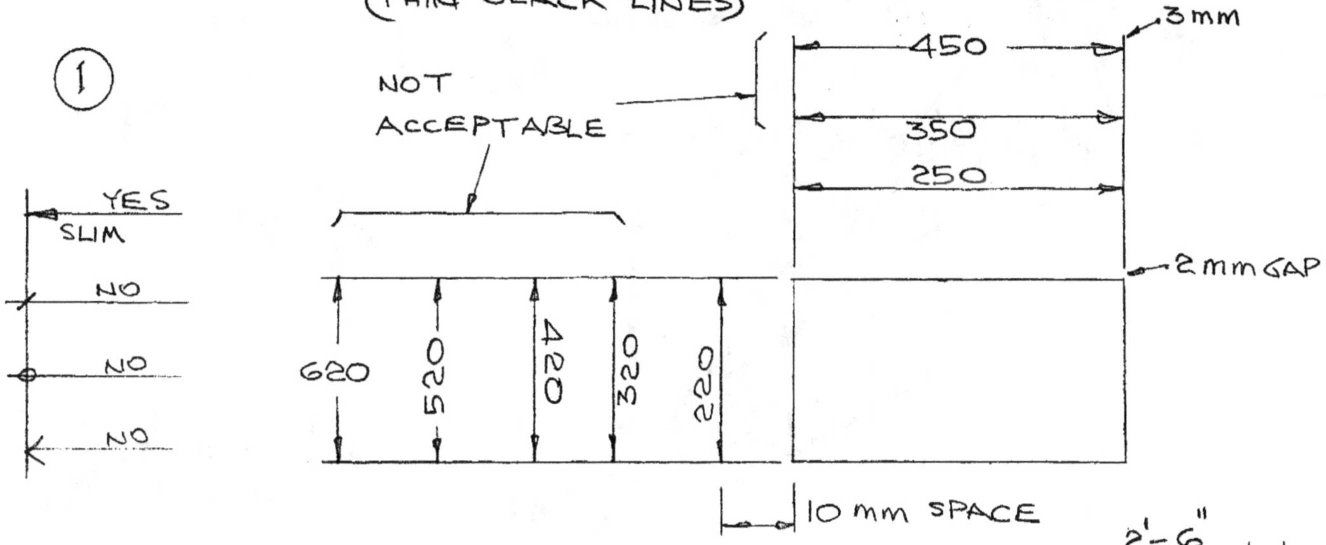

YES SLIM
NO
NO
NO

NOT ACCEPTABLE

450 — 3mm
350
250
— 2mm GAP

620 520 420 320 220

10 mm SPACE

2'-6" NARROW DIM.

② STRING DIM. & OVERALL DIM. ON SAME SIDE OF VIEW

KEEP STRING IN LINE

OVER-DIMENSIONING IS ACCEPTABLE (NOT IN MECH^{NL})

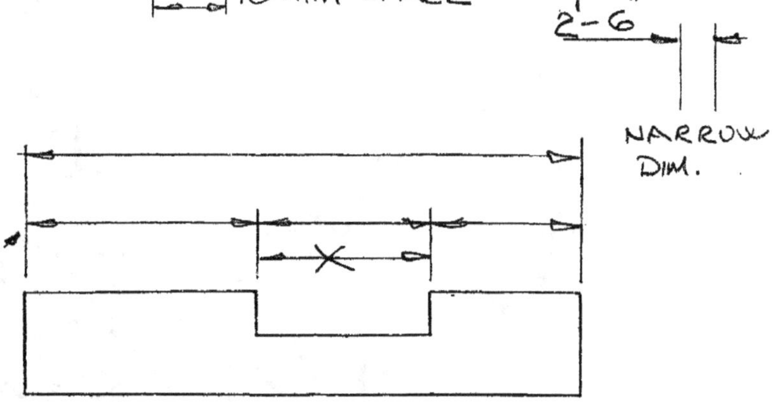

③

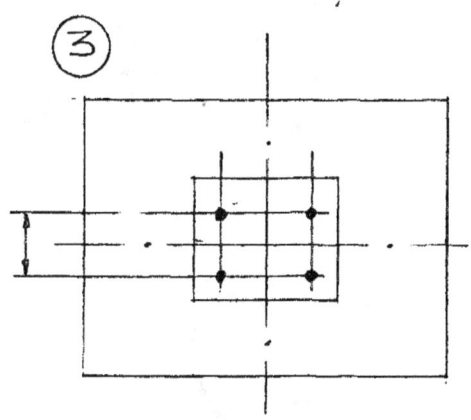

CENTRELINES: ARE CONTINUOUS
THIN LINES "DOT-DASH"
ANCHOR BOLTS ARE BLACK DOTS

④ UNITS: MILLIMETRES, NO DEC. PT., DO NOT SHOW "mm" ON DIMEN.

ACCEPTABLE TO USE "mm" IN NOTES:

e.g. 75 mm CLEARANCE
25 mm BOLT
50 mm PLATE
600 mm PIPE

ACCEPTABLE:

7	2"
65	1'-0"
432	2'-0½" ~~2'½"~~
6345	1'-1½"
25 000	

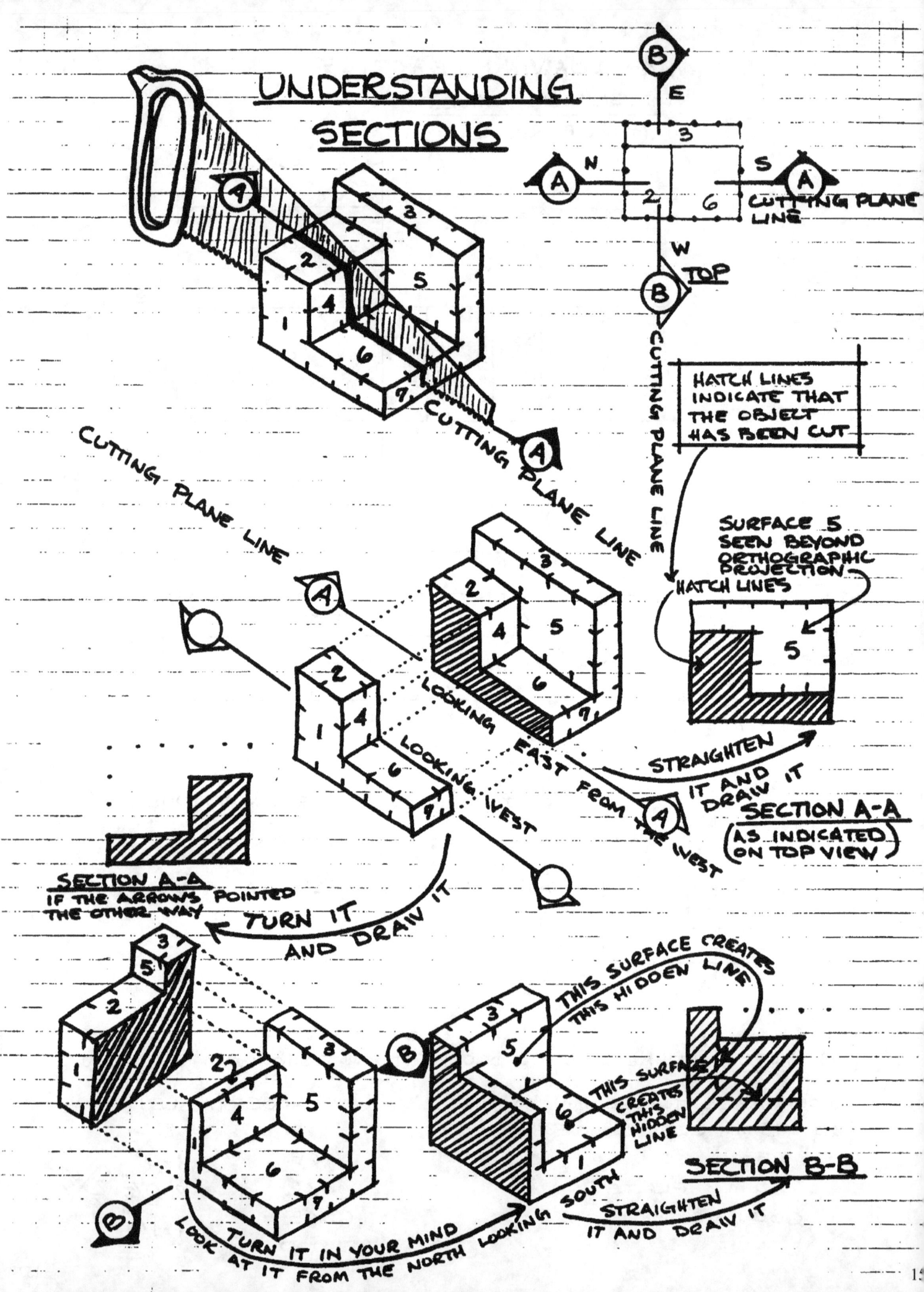

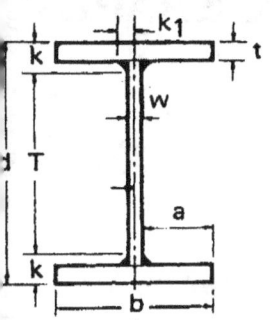

WELDED WIDE FLANGE SHAPES
WWF1200-WWF700

DIMENSIONS AND SURFACE AREAS

Nominal Mass	Depth d	Flange		Web Thickness w	Distances					Surface per metre	Designation ‡
		Width b	Thickness t		a	T	k	k_1	d-2t	Total	
kg/m	mm	mm	mm	mm	mm	mm	mm	mm	mm		
487	1 200	550	40.0	16.0	267	1 102	49	16	1 120	4.57	WWF1200 X487
403	1 200	550	30.0	16.0	267	1 122	39	16	1 140	4.57	X403
364	1 200	500	28.0	16.0	242	1 126	37	16	1 144	4.37	X364

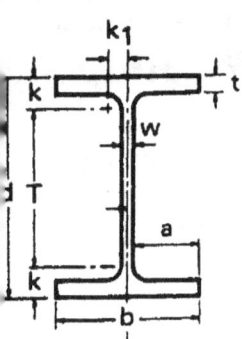

W SHAPES
W920 - W690

DIMENSIONS AND SURFACE AREAS

Nominal Mass	Theoretical Mass	Depth d	Flange			Web Thickness w		Distances					Surface per met	Designation ‡
			Width b	Mean Thickness t				a	T	k	k_1	d-2t	Total	
kg/m	kg/m	mm	mm	mm	mm	mm	mm	mm	mm	mm	mm	mm		
446	447.2	933	423	42.7	43	24.0	24	200	792	70	38	848	3.51	W920 X446*
417	418.1	928	422	39.9	40	22.5	22	200	793	67	37	848	3.50	X417*

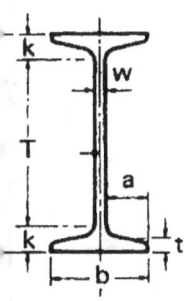

S SHAPES
S610 - S180

DIMENSIONS AND SURFACE AREAS

Nominal Mass	Theoretical Mass	Depth d	Flange			Web Thickness w		Distances			Surface per metre	Designation #
			Width b	Mean Thickness t				a	T	k	Total	
kg/m	kg/m	mm	mm	mm	mm	mm	mm	mm	mm	mm		
180	180.0	622	204	27.7	28	20.3	20	92	522	50	2.02	S610 X180*
158	157.8	622	200	27.7	28	15.7	16	92	522	50	2.01	X158*

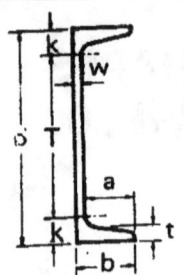

STANDARD CHANNELS (C SHAPES)

DIMENSIONS AND SURFACE AREAS

Nominal Mass	Theoretical Mass	Depth d	Flange		Web Thickness w	Distances			Surface per metre Total	Designation#
			Width b	Mean Thickness t		a	T	k		
kg/m	kg/m	mm	mm	mm	mm	mm	mm	mm	mm	
										C380
74	74.4	381	94	16.5	16	18.2	18	76	312	34	1.10	X74
60	59.4	381	89	16.5	16	13.2	13	76	312	34	1.09	X60
50	50.5	381	86	16.5	16	10.2	10	76	312	34	1.09	X50

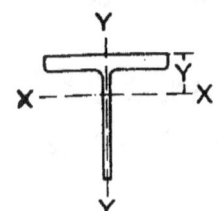

STRUCTURAL TEES
Cut from W Shapes

PROPERTIES AND DIMENSIONS

Designation	Dead Load	Total Area	Depth of Tee "d"	Flange		Stem Thickness "w"	Axis X-X				Axis Y-Y		
				Width	Aver. Thickness		I_x	S_x	r_x	y	I_y	S_y	r_y
	kN/m	mm²	mm	mm	mm	mm	10^6mm⁴	10^3mm³	mm	mm	10^6mm⁴	10^3mm³	mm
WT460													
X223*	2.18	28 500	466	423	42.7	24.0	510	1 410	134	105	270	1 280	97.3
X208.5*	2.04	26 600	464	422	39.9	22.5	474	1 310	133	103	250	1 190	96.9
X193.5*	1.89	24 700	460	420	36.6	21.3	439	1 230	133	103	226	1 080	95.7

ANGLES
Equal Legs

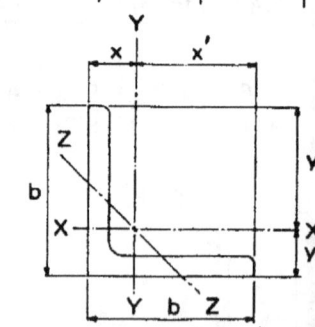

PROPERTIES AND DIMENSIONS

Size	Thickness t	Mass	Dead Load	Area	Axis X-X and Axis Y-Y				Axis Z-Z
					I	S	r	x or y	r
mm x mm	mm	kg/m	kN/m	mm²	10^6mm⁴	10^3mm³	mm	mm	mm
200X200									
	30	87.1	0.855	11 100	40.3	290	60.3	60.9	39.0
	25	73.6	0.722	9 380	34.8	247	60.9	59.2	39.1
	20	59.7	0.585	7 600	28.8	202	61.6	57.4	39.3
	16	48.2	0.473	6 140	23.7	165	62.1	55.9	39.5
	13	39.5	0.387	5 030	19.7	136	62.6	54.8	39.7
	10	30.6	0.300	3 900	15.5	106	63.0	53.7	39.9

WELDED WIDE FLANGE TEES
Cut from WWF Shapes

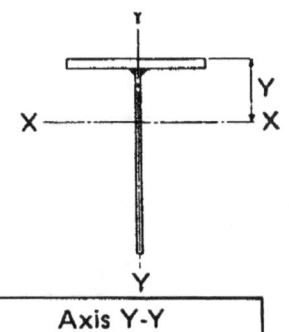

PROPERTIES AND DIMENSIONS

Designation	Dead Load	Total Area	Depth of Tee "d"	Flange Width	Flange Aver. Thickness	Stem Thickness "w"	Axis X-X I_x	Axis X-X S_x	Axis X-X r_x	Axis X-X y	Axis Y-Y I_y	Axis Y-Y S_y	Axis Y-Y r_y
	kN/m	mm²	mm	mm	mm	mm	10^6mm⁴	10^3mm³	mm	mm	10^6mm⁴	10^3mm³	mm
WWT275													
X360.5	3.53	46 000	275	550	60.0	60.0	235	1 140	71.5	68.6	836	3 040	135
X310	3.04	39 500	275	550	60.0	30.0	137	615	58.9	52.5	832	3 030	145
X251.5	2.46	32 100	275	550	50.0	20.0	97.8	424	55.2	44.4	693	2 520	147
X210	2.05	26 800	275	550	40.0	20.0	97.8	424	60.4	44.2	555	2 020	144
X108.5	1.06	13 900	275	550	20.0	11.0	57.8	244	64.5	37.9	277	1 010	141

HOLLOW STRUCTURAL SECTIONS
Square

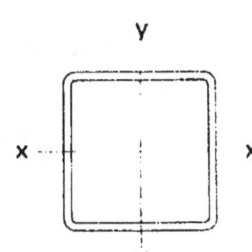

PROPERTIES AND DIMENSIONS

Outside Dimensions	Wall Thickness	Mass	Dead Load	Area	I	S	r	Z	Torsional Constant J	Surface Area	Shear Constant C_{rt}
mm×mm	mm	kg/m	kN/m	mm²	10^6mm⁴	10^3mm³	mm	10^3mm³	10^3mm⁴	m²/m	mm²
304.8 X 304.8	12.70	113	1.11	14 400	202	1 330	118	1 560	324 000	1.18	6 450
	11.13	100	0.982	12 800	181	1 190	119	1 390	288 000	1.18	5 790
	9.53	86.5	0.848	11 000	158	1 040	120	1 210	250 000	1.19	5 080
	7.95	72.8	0.714	9 280	135	886	121	1 030	211 000	1.19	4 340
	6.35	58.7	0.576	7 480	110	723	121	833	171 000	1.20	3 550

HOLLOW STRUCTURAL SECTIONS
Round

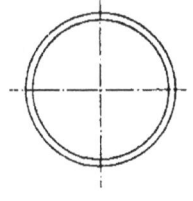

PROPERTIES AND DIMENSIONS

Outside Diameter	Wall Thickness	Mass	Dead Load	Area	I	S	r	Z	J	Surface Area	Shear Constant C_{rt}
mm	mm	kg/m	kN/m	mm²	10^6mm⁴	10^3mm³	mm	10^3mm³	10^3mm⁴	m²/m	mm²
406.4	12.70	123	1.21	15 700	305	1 500	139	1 970	609 000	1.28	7 860
	11.13	108	1.06	13 800	270	1 330	140	1 740	540 000	1.28	6 910
	9.53	93.3	0.915	11 900	234	1 150	140	1 500	468 000	1.28	5 940
	7.95	78.1	0.766	9 950	198	972	141	1 260	395 000	1.28	4 970
	6.35	62.6	0.614	7 980	160	786	141	1 020	319 000	1.28	3 990

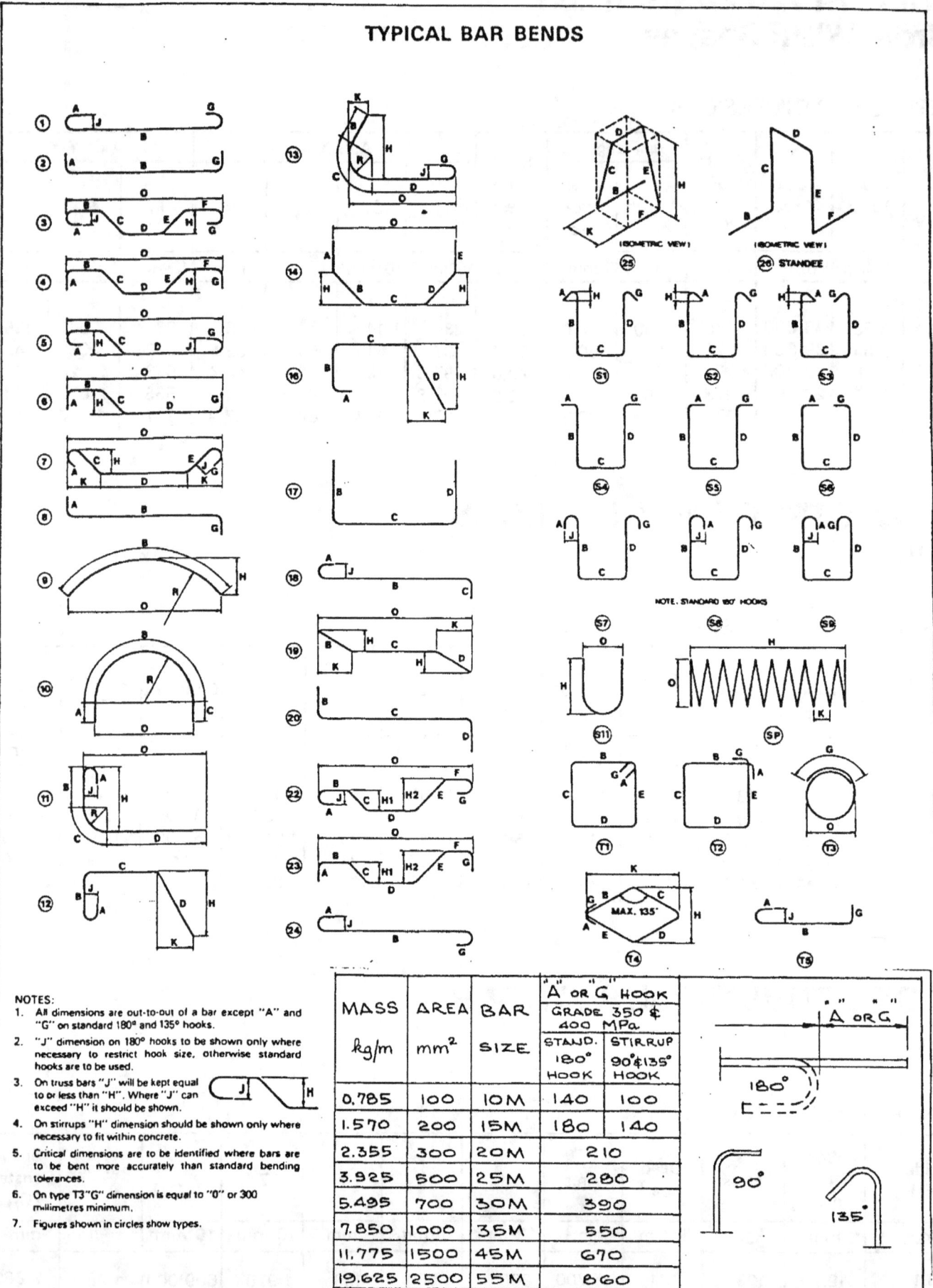

Aban	Abandon(ed)	CLF	Chain Link Fence
Abut	Abutment	Chkd	Checked
Allce	Allowance	Circ	Circular
Asph	Asphalt	CS	Circular Curve to Spiral
AC	Asphalt Coated; Asphalt Curb or Asphalt Cement	CIP	Clay Pipe
		Con	Concession
AC PI	Asphalt Coated and Paved Invert	Conc or C	Concrete
AC & G	Asphalt Curb and Gutter	CC	Concrete Curb
AG	Asphalt Gutter	CC & G	Concrete Curb and Gutter
A S/W	Asphalt Sidewalk	CG	Concrete Gutter
Ave	Avenue	CP	Concrete Pipe
		C S/W	Concrete Sidewalk
		Const	Construction
		Cont	Contract
Bfl	Backfill	CAH	Controlled Access Highway
BS	Beginning of Superelevation or Bottom of Slope	CSP	Corrugated Steel Pipe
		CSPA	Corrugated Steel Pipe Arch
BWF	Barbed Wire Fence	Co	County
B/L	Base Line	Cr	Creek or Crushed
Brg	Bearing(s)	Culv	Culvert
BC	Beginning of Curve		
BFS	Beginning of Full Superelevation		
BVC	Beginning of Vertical Curve		
BCA	Bell Cable Aerial	Def	Deflection
BCU	Bell Cable Underground	Deg	Degree(s)
BCUM	Bell Cable Underground Marker	Dia	Diameter
BM	Bench Mark	Dist	District
Blk	Block	D	Ditch
BF	Board Fence	DI	Ditch Inlet or Ductile Iron
Bld(y)	Boulder(y)	DICB	Ditch Inlet Catch Basin
Blvd	Boulevard	Div	Division
BBGR	Box Beam Guide Rail	Dwg	Drawing
Br	Brick	D/W	Driveway
Bldg(s)	Building(s)		
CGR	Cable Guide Rail	EC	Earth Cut or End of Curve
CSA	Canadian Standards Association	EF	Earth Fill or Electric Fence
CI	Cast Iron	ED	Earth Ditching
CB	Catch Basin	EB	Eastbound
CBMH	Catch Basin – Manhole	EBL	Eastbound Lane
C/C	Centre to Centre	EP	Edge of Pavement
C/L or ℄	Centre Line	ES	Edge of Shoulder

ONTARIO PROVINCIAL STANDARD DRAWING	Date	1986 07 18	Rev	1

ABBREVIATIONS

Date

OPSD-100.01

EMH	Electrical Manhole	Hwy	Highway
Elev	Elevation	Hor	Horizontal
EFS	End of Full Superelevation	HL	Hot Laid
ES	End of Superelevation	HM	Hot Mix
EVC	End of Vertical Curve	HOC	Hub on Curve
Ent	Entrance	HOST	Hub on Subtangent
Exc	Excavation	HOT	Hub on Tangent
Expy	Expressway	Hyd	Hydrant
Extn	Extension		
E_T	External Distance of a Spiralled Curve		
		Incl	Include (ing) (sive)
		ID	Inside Diameter
FL	Fence Line	IF	Inside Face
Ftg	Footing	Inst	Instrument
Fdn	Foundation	Inv Elev	Invert Elevation
Fr	Frame	IB	Iron Bar
		IF	Iron Fence
		IP	Iron Pipe
Galv	Galvanized		
Ga	Gauge		
GP	Grade Point	Jt	Joint
G	Grading		
GC	Grading and Culverts		
GD	Grading and Drainage		
Gran	Granular		
GBfl	Granular Backfill	L	Length or Length of Curve
GB	Granular Base	La	Length of Spiral between Circular Curve
GBC	Granular Base Course		
GSB	Granular Sub Base	Ls	Length of Spiral Curve
Gr	Gravel (ly)	LVC	Length of Vertical Curve
GL	Ground Level	LT	Long Tangent
		LWL	Low Water Level
Hw	Headwall		
Ht	Height	MH	Manhole
HT	High Tension	Matl	Material
HTL	High Tension Line	Max	Maximum
HWL	High Water Level	Med	Median
HWM	High Water Mark	Min	Minimum

ONTARIO PROVINCIAL STANDARD DRAWING Date 1986 04 07 Rev 1

ABBREVIATIONS

OPSD-100.02

MTC	Ministry of Transportation and Communications Ontario	R	Radius
		RF	Rail Fence or Rock Fill
Mon	Monument	Rwy	Railway
Mun	Municipal (ity)	RP	Reference Point or Registered Plan
		Reg	Region (al)
		Reinf	Reinforced
		Reqd	Required
NC	Normal Crown	Resurf	Resurfacing
N	North (ing)	RW	Retaining Wall
NB	Northbound	Rev	Revision
NBL	Northbound Lane	ROW	Right of Way
		RS	Ripple Strip
		RR	Rip-Rap (all types)
		Rk	Rock
		RC	Rock Cut
		R Exc	Rock Excavation
OC	On Centre	Rnd	Rounding
Org M	Organic Matter		
Orig	Original		
OG	Original Ground		
OD	Outside Diameter		
OF	Outside Face		
Opass	Overpass	Sa	Sand
OPSD	Ontario Provincial Standard Drawing	San	Sanitary
		Sec	Second (s) (ary)
OPSS	Ontario Provincial Standard Specification	Serv Rd	Service Road
		Sh	Shatter or shoulder
		Sh Rk	Shot Rock
		SD	Side Ditch
		SR	Sideroad
		S/W	Sidewalk
Pavt	Pavement	S	Superelevation, Rate of
Perf	Perforated	SB	Southbound
PF	Picket Fence	SBL	Southbound Lane
PG	Plain Galvanized or Profile Grade	SP	Special Provision
PI	Point of Intersection	Spec	Specification
PRC	Point of Reverse Curve	SC	Spiral to Circular Curve
PVT	Point on Vertical Tangent	ST	Spiral to Tangent or Short Tangent
PCC	Point of Compound Curve	Std	Standard
POC	Point on Curve	Sta	Station
PVC	Point on Vertical Curve or Polyvinyl Chloride	SBGR	Steel Beam Guide Rail
		SF	Stone Fence
Prof	Profile	Sty	Storey
PC	Profile Control	Stm	Storm
PL	Property Line	St	Stripping
Prop	Proposed		

ONTARIO PROVINCIAL STANDARD DRAWING

ABBREVIATIONS

Date 1983 12 01

OPSD-100.03

Str	Structure (al)	WL	Water Level
SPCSA	Structural Plate Corrugated Steel Arch (with footings)	WT	Water Table
		WB	Westbound
SPCSP	Structural Plate Corrugated Steel Pipe	WBL	Westbound Lane
SPCSPA	Structural Plate Corrugated Steel Pipe Arch	WW	Wing Wall
		WF	Wire Fence
Surf	Surface	WMF	Wire Mesh Fence
		WP	Work Project
		WIF	Wrought Iron Fence

TS	Tangent to Spiral or Tensile Strength
TB	Top of Bank or Turning Basin
TG or T/G	Top of Grate
TP	Top of Pavement
TR or T/R	Top of Rail
Twp	Township
TCH	Trans Canada Highway
Trans	Transverse or Transferred
Typ	Typical

METRIC UNITS

cm	Centimetre
°C	Degree Celsius
ha	Hectare
kg	Kilogram(s)
km	Kilometre(s)
kN	Kilonewton(s)
kPa	Kilopascal(s)
L	Litre(s)
m	Metre(s)
m^2	Square metre
m^3	Cubic metre
mm	Millimetre(s)
MPa	Megapascal(s)
t	Tonne

Upass	Underpass

Vert	Vertical
VC	Vertical Curve
VCP	Vitrified Clay Pipe
Vol	Volume

ONTARIO PROVINCIAL STANDARD DRAWING

Date: 1983 12 01 | Rev

ABBREVIATIONS

OPSD-100.04

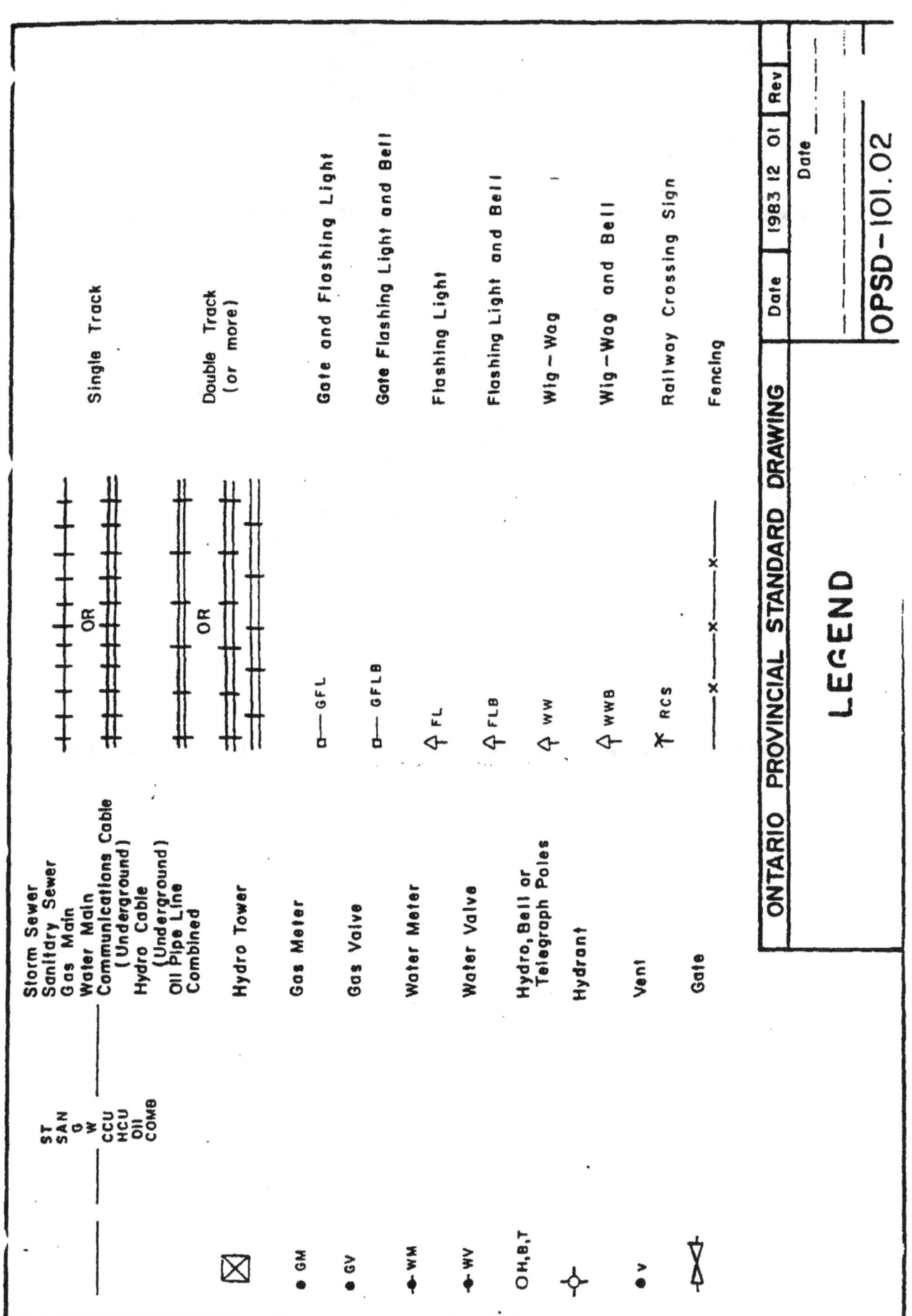

Department of
BUILDING AND CONSTRUCTION SCIENCES
Mohawk College

EXAMPLES OF PREVIOUS STUDENT WORK

FOR: MISS D. SARNA
FROM: GEORGE VIDOVIC TCA22

MOHAWK COLLEGE OF APPLIED ARTS AND TECHNOLOGY
COURSE OUTLINE

A. IDENTIFICATION ITEMS

COURSE NAME   DRAFTING 2
PROGRAM NAME   ARCHITECTURAL CIVIL TRANSPORT
DURATION   TOTAL HRS 56 TOTAL WKS 14
DEPARTMENT   BUILDING & CONSTRUCTION SCIENCE
FACULTY   FACULTY OF ENGINEERING TECH

GENERAL COURSE OBJECTIVE

BY THE COMPLETION OF THIS COURSE THE STUDENT WILL HAVE FURTHER DEVELOPED LOGIC THINKING AND DESIGN SKILLS IN THE FIELDS OF ARCHITECTURE CIVIL TRANSPORTATION AND CONSTRUCTION. STUDENTS WILL PREPARE PRELIMINARY RESIDENTIAL DESIGN DRAWINGS FOR CLIENT APPROVAL FOLLOWED BY A SET OF DETAILED CONSTRUCTION DRAWINGS. TO SENSITIZE STUDENTS TO ENVIRONMENTAL CONSIDERATIONS BASIC ELEMENTS OF PASSIVE SOLAR AND ENERGY CONSERVATION WILL BE INCORPORATED IN THE BUILDING DESIGN. STUDENTS WILL PREPARE WORKING DRAWINGS FOR STRUCTURAL STEEL FRAMEWORK CONCRETE WORKS FOOTINGS AND REINFORCING STEEL. STUDENTS WILL ALSO PREPARE WORKING DRAWINGS PLANS CROSS-SECTIONS AND PROFILES OF ROADWAYS AND OFF-ROAD TRANSPORTATION FACILITIES.

REQUIRED TEXTBOOKS

1. COURSE TEXTBOOK "DRAFTING 2 MANUAL EA206" MOHAWK COLLEGE OCTOBER 1994
2. "ILLUSTRATED RESIDENTIAL AND COMMERCIAL CONSTRUCTION" PETER A. MANN PRENTICE HALL INC. 1989

ADDTIONAL REFERENCES/BIBLIOGRAPHY

1. NATIONAL AND O.B.C. CODE BOOKS.
2. TECHNICAL REFERENCE BOOKS AS DIRECTED BY THE INSTRUCTOR.
3. SAMPLE RESIDENTIAL DRAWING SETS.
4. SAMPLE STRUCTURAL STEEL OR CONCRETE DRAWING SETS.
5. SAMPLE TRANSPORTATION DRAWING SETS.

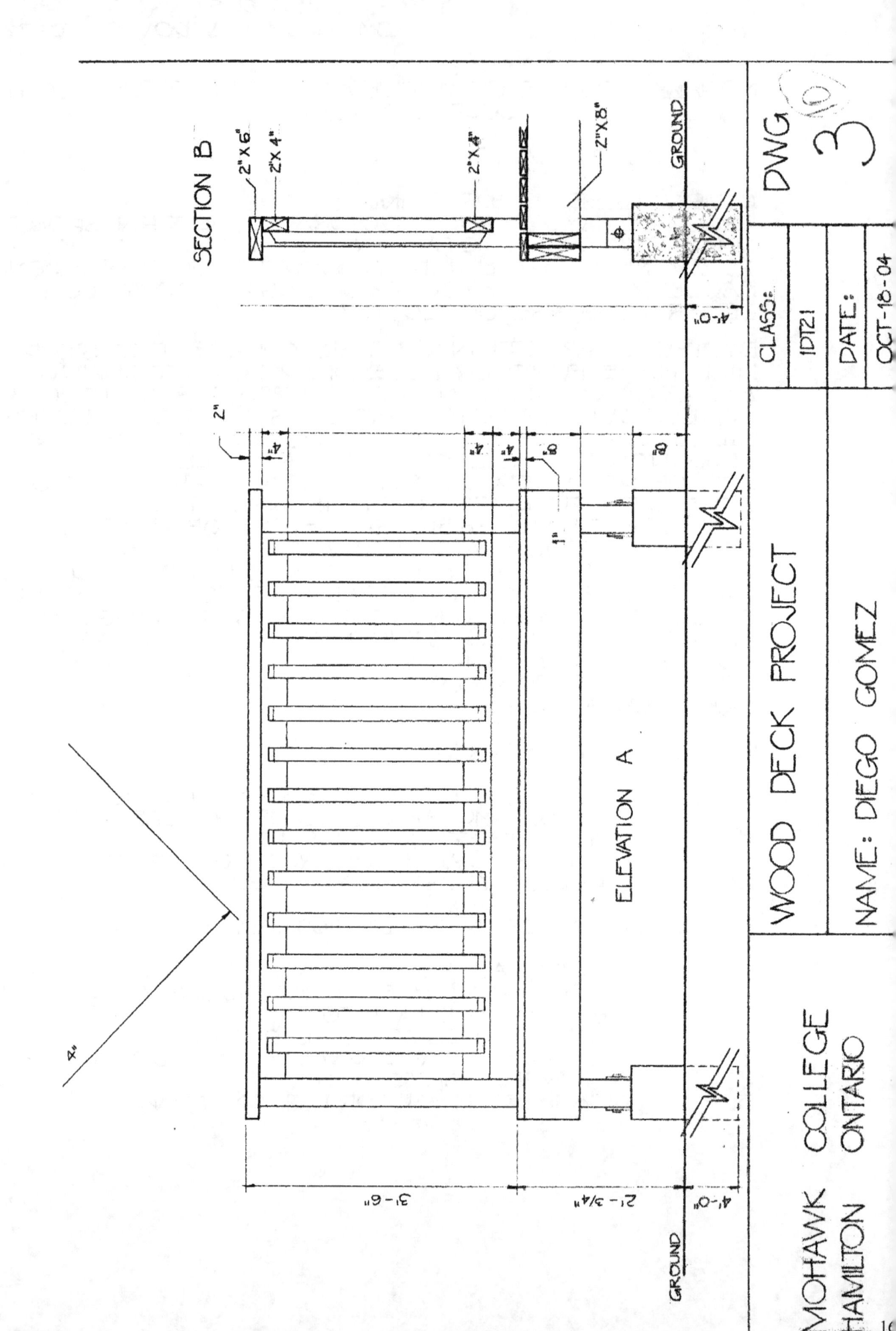

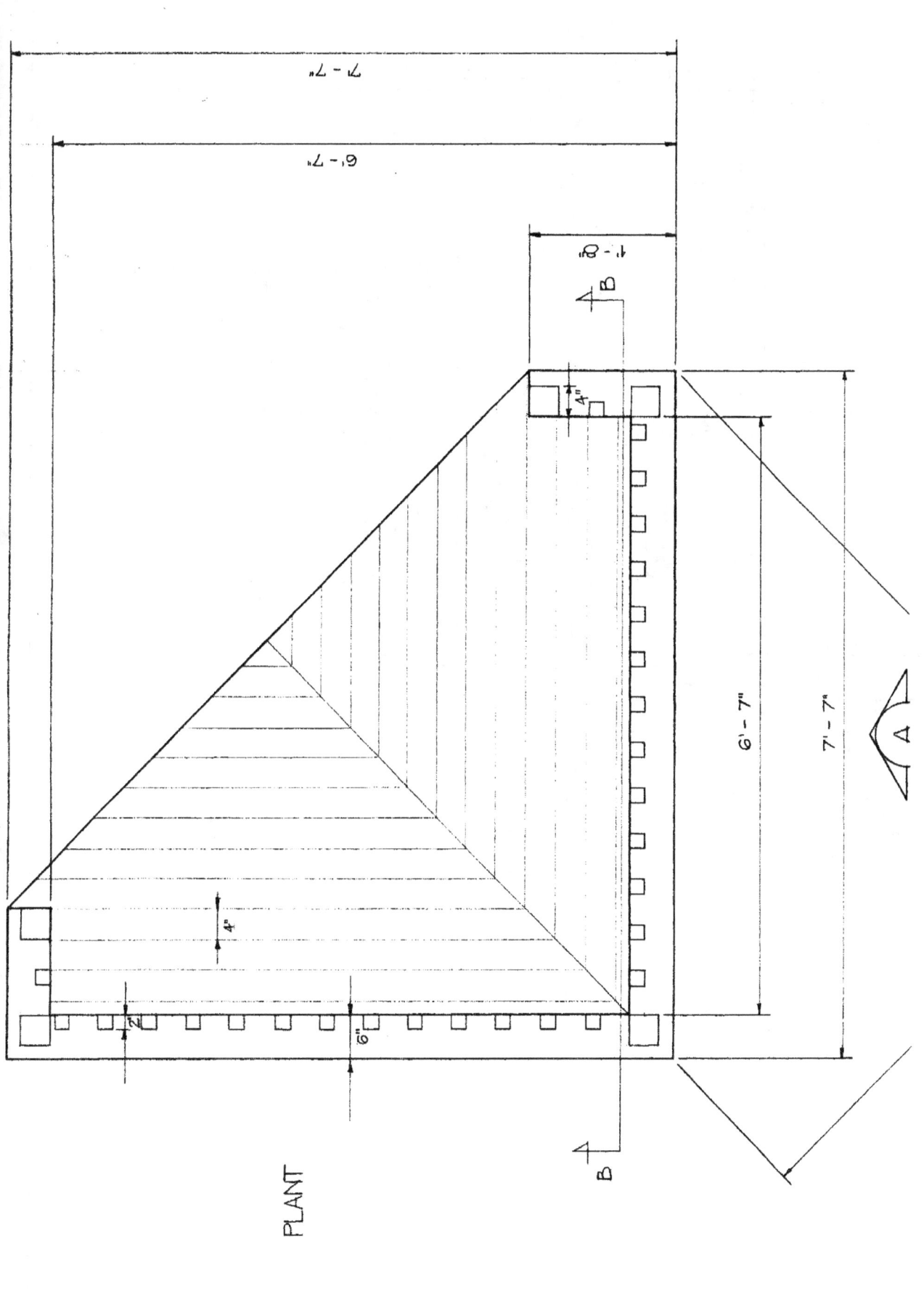

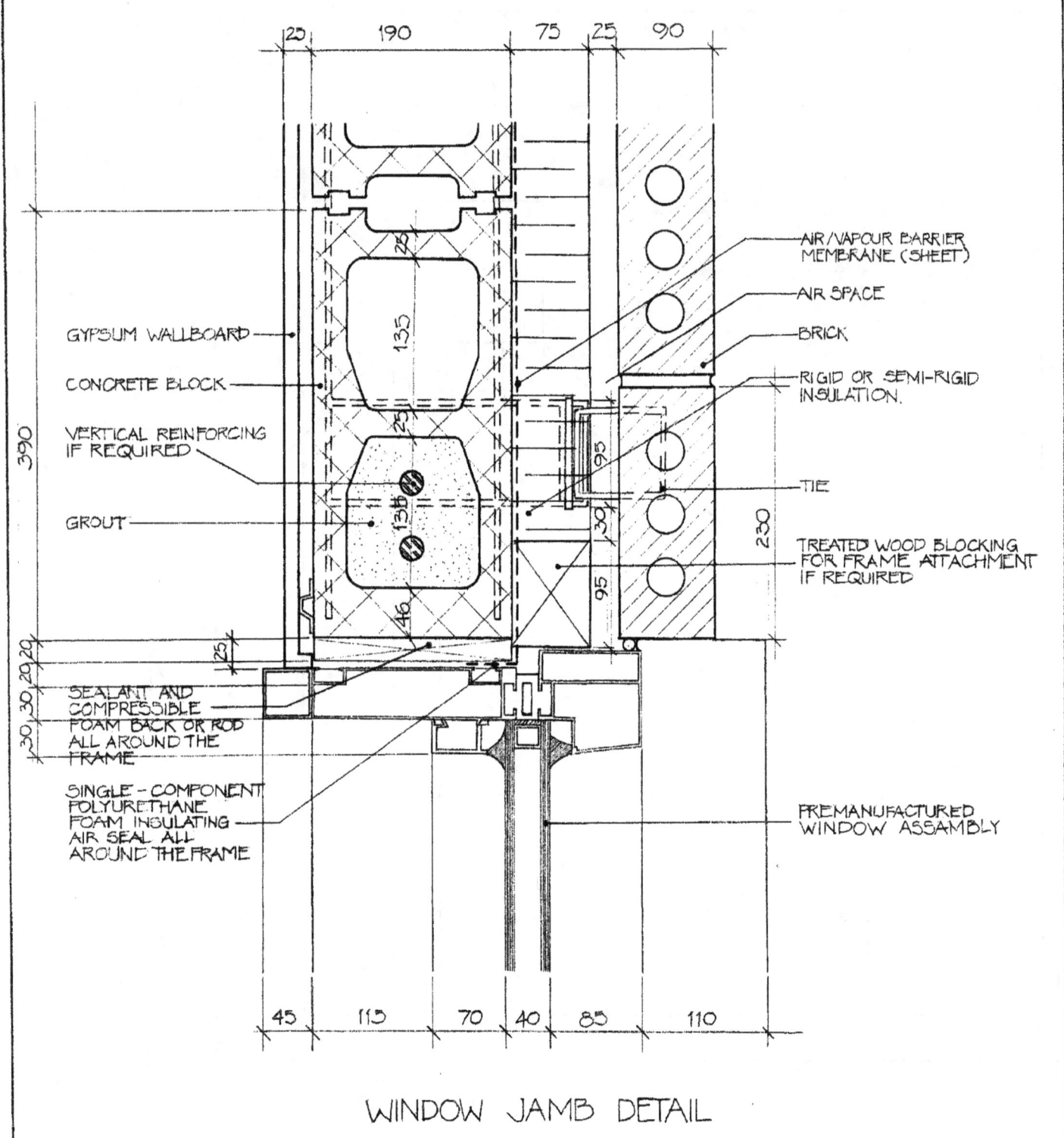

WINDOW JAMB DETAIL

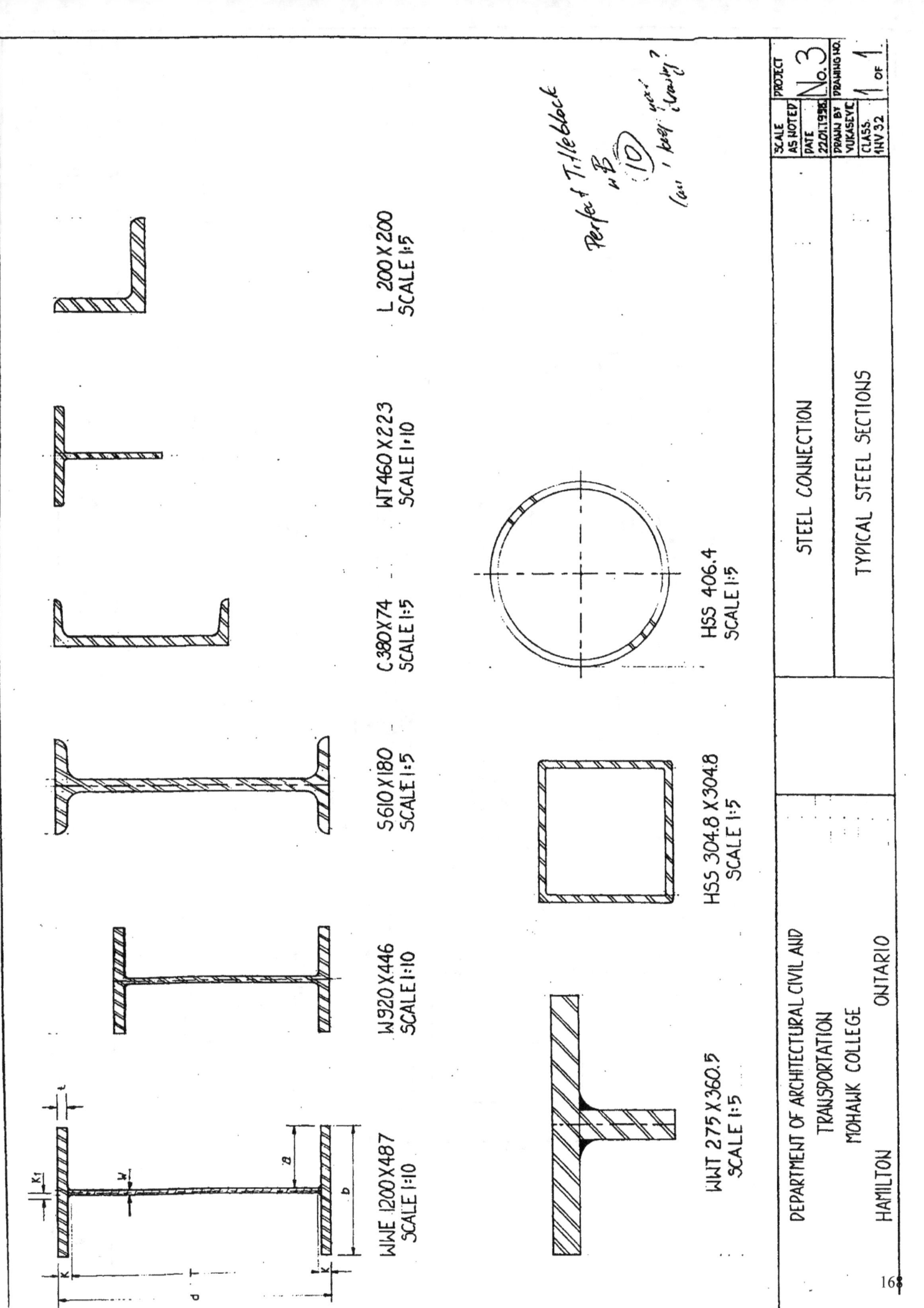

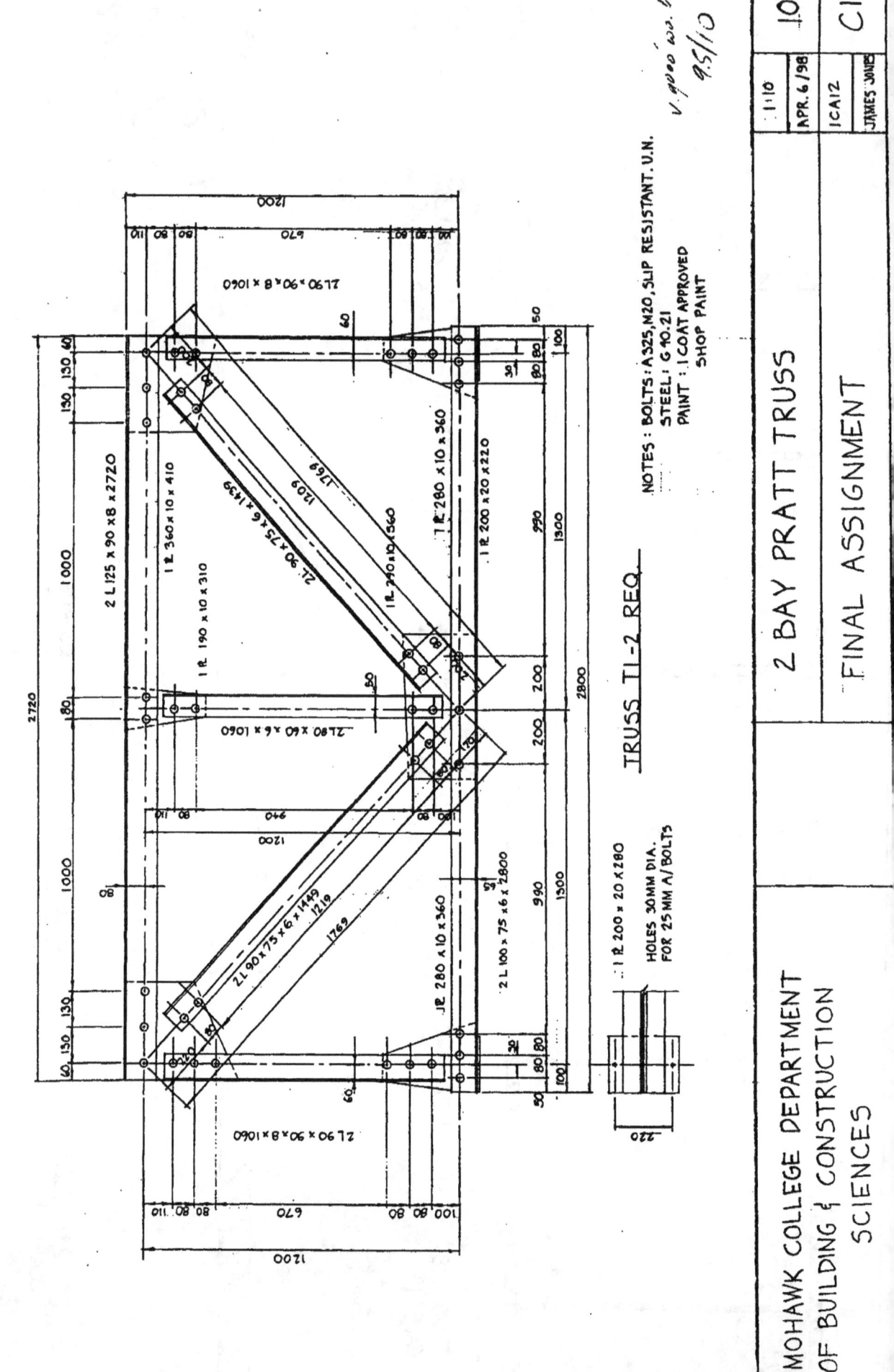

Department of
BUILDING AND CONSTRUCTION SCIENCES
Mohawk College

EXAMPLES OF PROFESSIONAL WORK

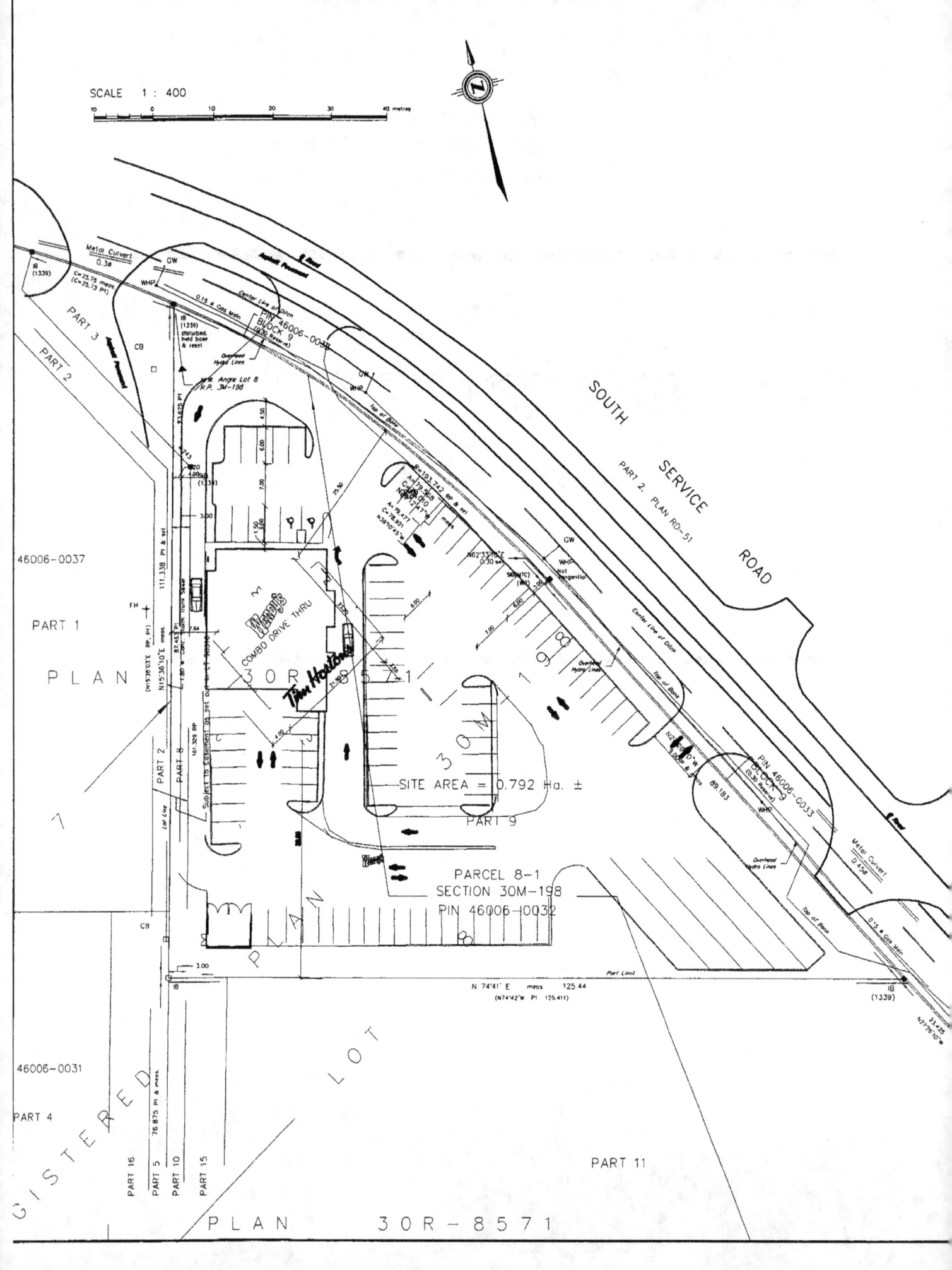

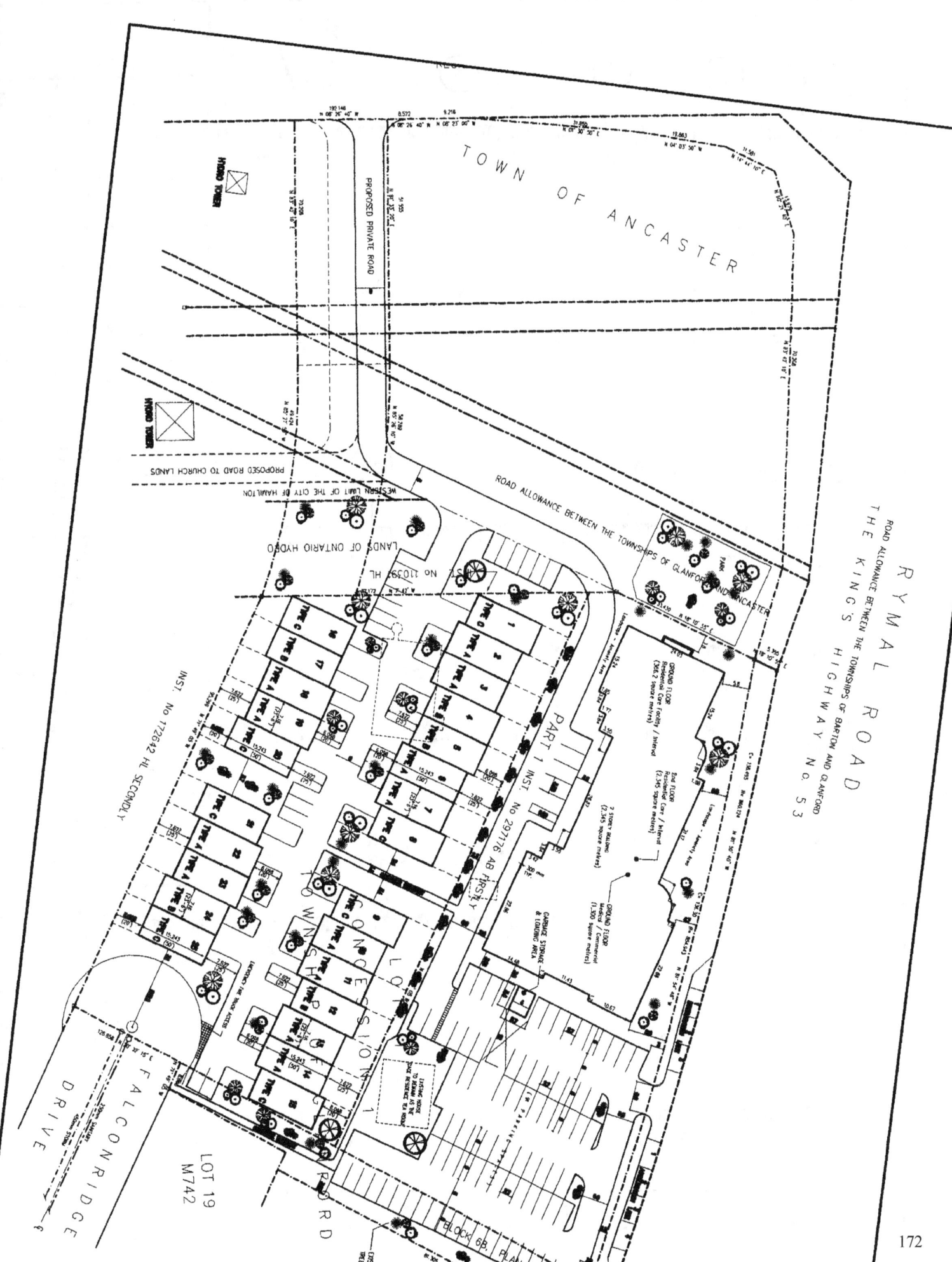

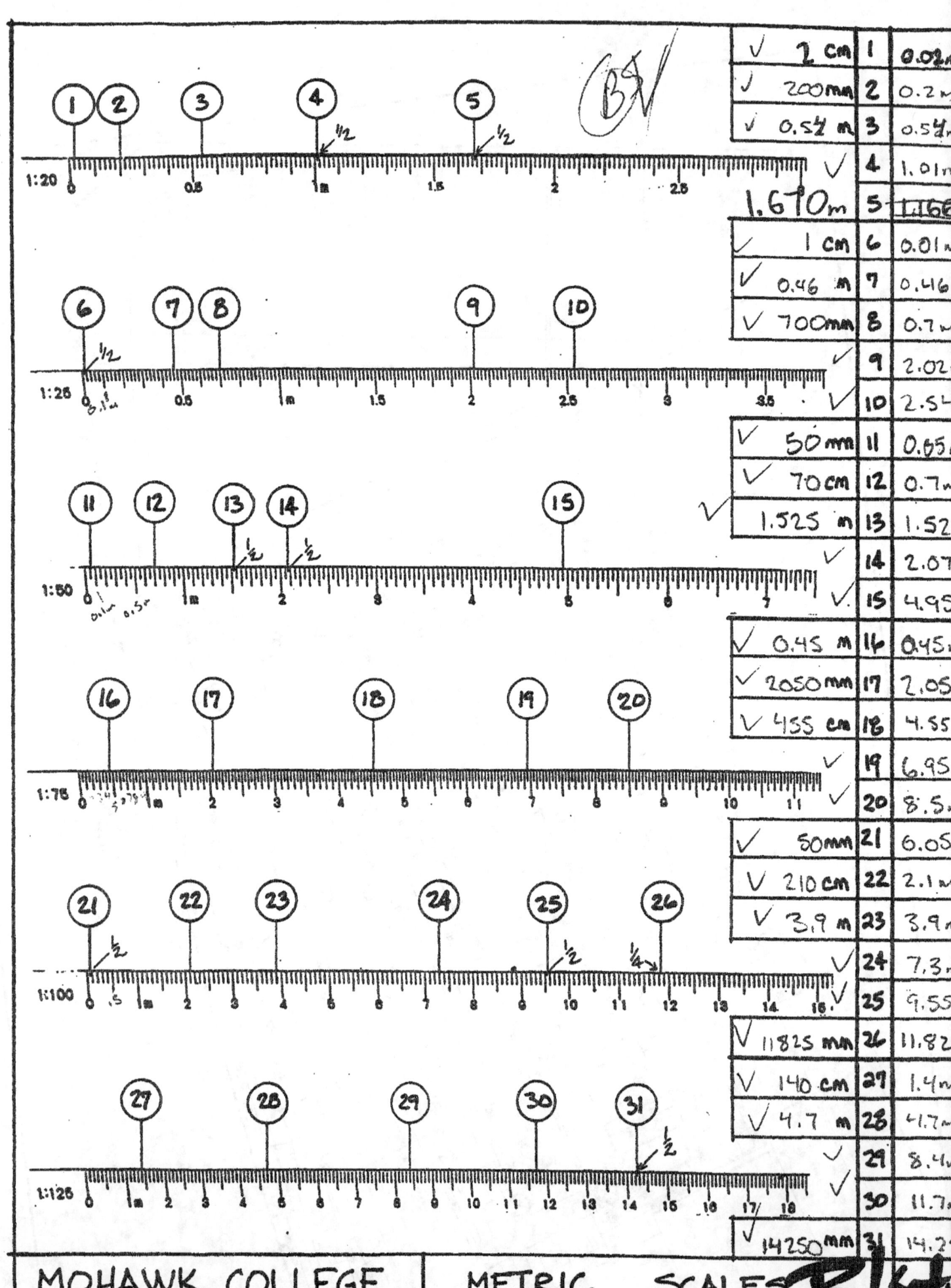

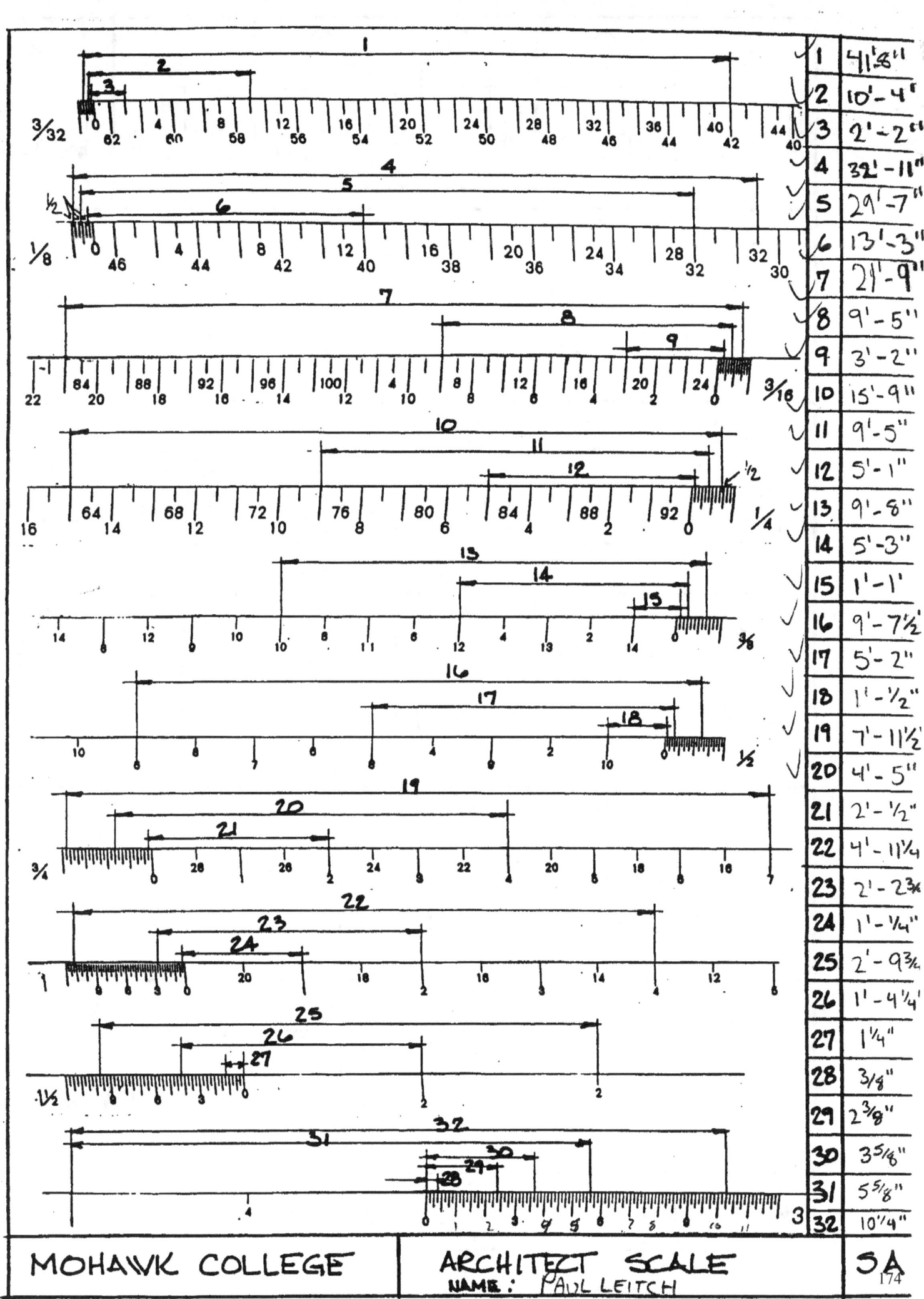

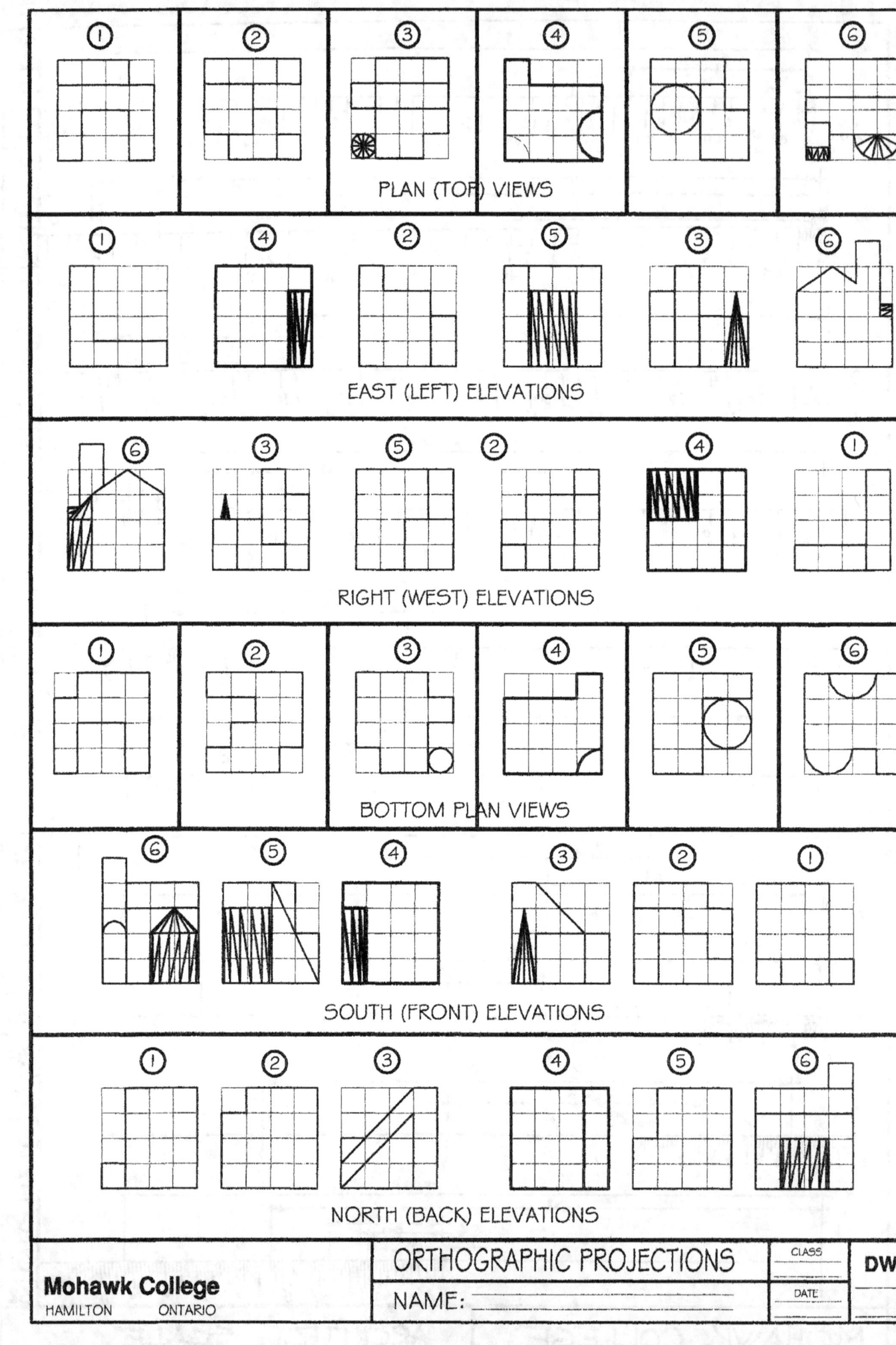

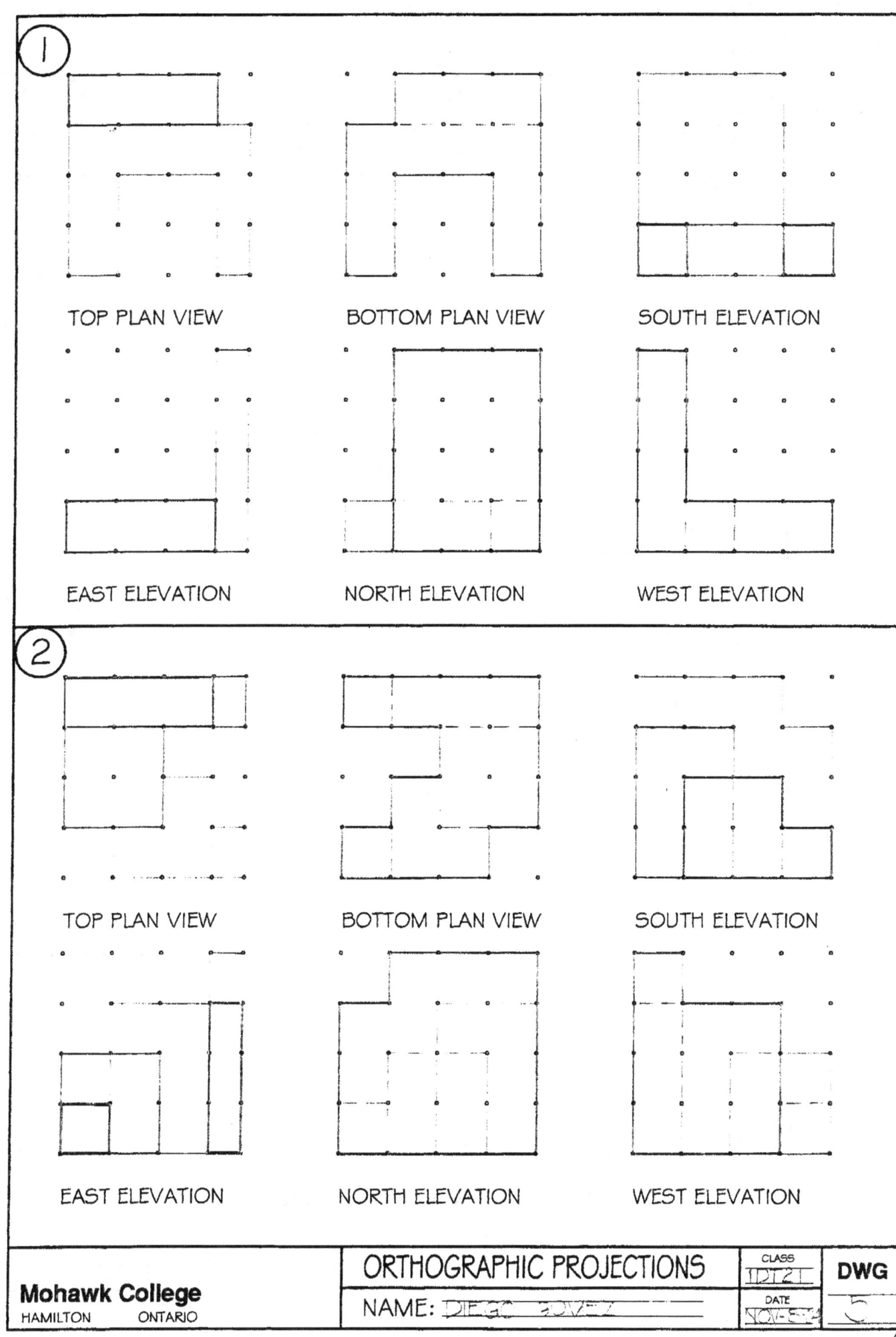

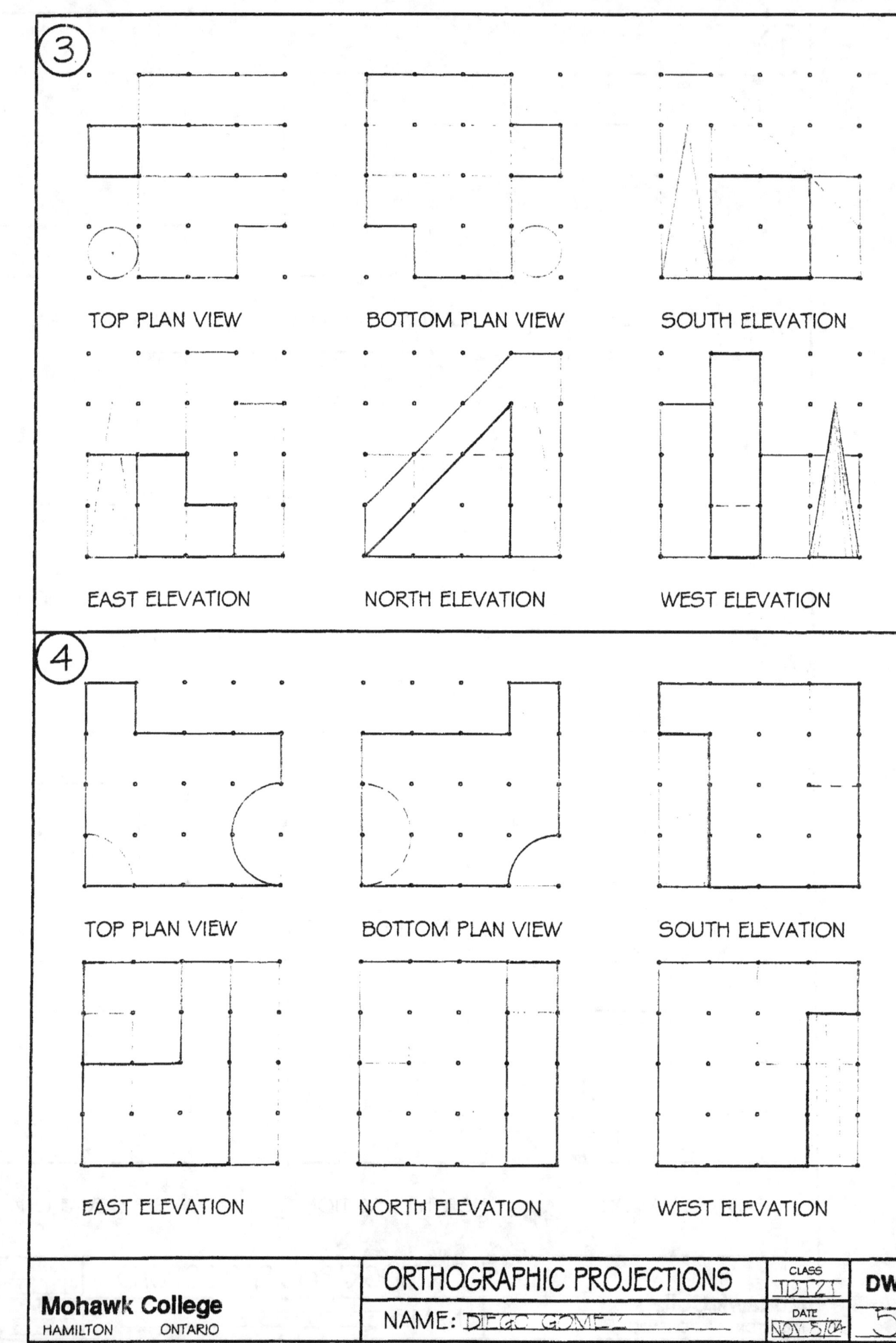

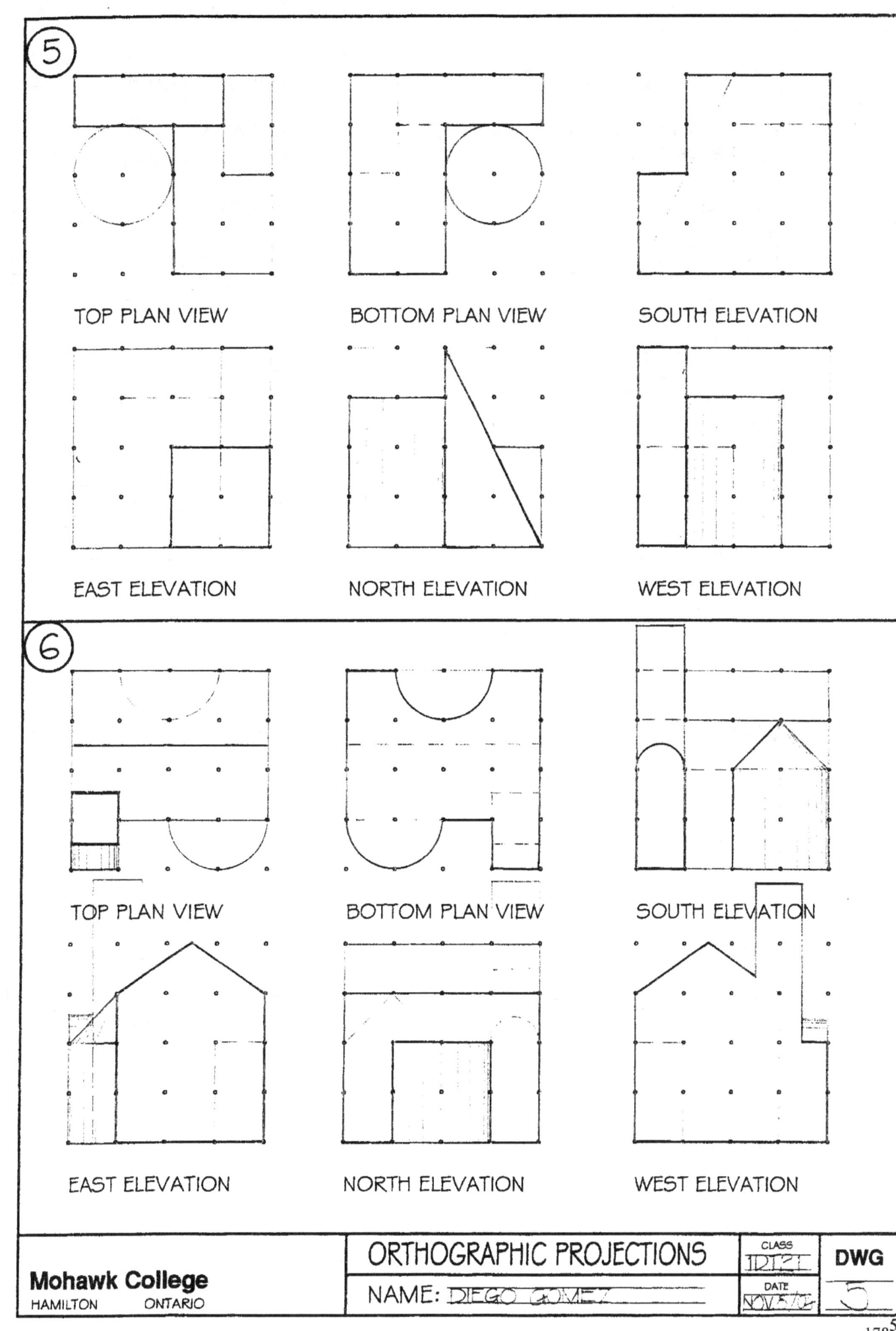

DRAFTING EVALUATION SHEET

You lost marks on your drawing due to the following:

LINEWORK	
PRINTING	
DIMENSIONING	
TITLE BLOCK	
NEATNESS	
ARROWHEADS	
LACK OF INFORMATION	
WRONG INFORMATION	
LAYOUT	
LATE	

TOTAL OUT OF 10	

Comments: _____
_____
_____

DRAFTING EVALUATION SHEET

You lost marks on your drawing due to the following:

LINEWORK	
PRINTING	
DIMENSIONING	
TITLE BLOCK	
NEATNESS	
ARROWHEADS	
LACK OF INFORMATION	
WRONG INFORMATION	
LAYOUT	
LATE	

TOTAL OUT OF 10	

Comments: _____
_____
_____

DRAFTING EVALUATION SHEET

You lost marks on your drawing due to the following:

LINEWORK	
PRINTING	
DIMENSIONING	
TITLE BLOCK	
NEATNESS	
ARROWHEADS	
LACK OF INFORMATION	
WRONG INFORMATION	
LAYOUT	
LATE	

TOTAL OUT OF 10	

Comments: _____
_____

DRAFTING EVALUATION SHEET

You lost marks on your drawing due to the following:

LINEWORK	
PRINTING	
DIMENSIONING	
TITLE BLOCK	
NEATNESS	
ARROWHEADS	
LACK OF INFORMATION	
WRONG INFORMATION	
LAYOUT	
LATE	

TOTAL OUT OF 10	

Comments: _____
_____

www.ingramcontent.com/pod-product-compliance
Lightning Source LLC
Chambersburg PA
CBHW080912170526
45158CB00008B/2079